THE ATTRACTION OF THE HIMALAYA MOUNTAINS

THE ATTRACTION OF THE HIMALAYA MOUNTAINS: UPON THE PLUMB-LINE IN INDIA

SIDNEY GERALD BURRARD

GYAN

Published by

Gyan Publishing House
5, Ansari Road
Daryaganj, New Delhi-110002
Phone: 011-47034999, 9811692060
E-mail: books@gyanbooks.com

Distribution Network
gyanbooks.com
India, USA, Canada, UK, Australia

ISBN: 978-93-7133-416-7 (HB)

First Published, 1901

2nd Impression 2026

Printed at: Gyan Press, Delhi.

THE ATTRACTION OF THE HIMALAYA MOUNTAINS
Author: SIDNEY GERALD BURRARD

Survey of India Department.

PROFESSIONAL PAPER—NO. 5.

THE ATTRACTION OF THE HIMALAYA MOUNTAINS UPON THE PLUMB-LINE IN INDIA.

CONSIDERATIONS OF RECENT DATA

BY

MAJOR S. G. BURRARD, ROYAL ENGINEERS,
SUPERINTENDENT TRIGONOMETRICAL SURVEYS.

PUBLISHED BY DIRECTION OF

COLONEL St. G. C. GORE, ROYAL ENGINEERS,
SURVEYOR GENERAL OF INDIA.

Dehra Dun:

PRINTED AT THE OFFICE OF THE TRIGONOMETRICAL BRANCH, SURVEY OF INDIA.

1901.

CONTENTS.

DIAGRAMS AND CHARTS.

EPITOME

OF THE INVESTIGATION.

DATE			PAGE.
1895	(1)	In a paper read before the Royal Society, General Walker advocates the employment of "groups" of astronomical stations for the purpose of eliminating the effects of local attraction. In the same paper he explains the apparent preponderance of northerly deflections throughout India by assuming, that local attraction is producing a *southerly* deflection at Kaliánpur, the station of reference of the Indian Survey.	3-4
1898	(2)	It is decided to determine the effect of local attraction at Kaliánpur by means of a "group" of astronomical stations.	
1898	(3)	Captain Lenox Conyngham commences to observe for both latitude and azimuth at all stations of the Kaliánpur group.	
1899	(4)	Completion of astronomical observations at Kaliánpur and at eight adjacent stations, four being situated at an average distance of 9 miles, and four at an average distance of 30 miles from Kaliánpur.	
February, 1900	(5)	An analysis of the results of the observations of the "group" shows that *local* attraction is producing a *northerly* deflection at Kaliánpur.	5-10
	(6)	Necessity arises of reconsidering General Walker's theory explaining the preponderance of *northerly* deflections throughout India.	10-14
June, 1900	(7)	An examination of data discloses the fact that two-thirds of the *southerly* deflections observed in India are to be found situated within a narrow zone, running from east to west between the parallels of 24° and 26°.	14
	(8)	Further examinations show that large *northerly* deflections prevail from east to west between the parallels of 24° and 18°.	21 and 102
	(9)	Great significance is attached to the fact that the parallel of 24°, along which the deflections change sign, happens to be the parallel of the station of reference of the Survey.	14-21

DATE			PAGE.
	(10)	It is assumed that the change in the sign of the deflections along the parallel of 24° is in some way connected with the relationship of this parallel to the station of reference.	14-21
	(11)	The observed latitudes in Sub-Himalayan regions preclude the acceptance of Pratt's theory, that the Himalayas are wholly compensated. If a large Himalayan attraction exists at Dehra Dún, it cannot suddenly cease south of Dehra Dún.	44-45
	(12)	It is suggested that, if the Himalaya Mountains and Indian Ocean are uncompensated and are influencing the plumb-lines throughout India, a change in the sign of the deflections would appear along the parallel of the station of reference.	14-21
	(13)	The well-known discrepancy between the results of calculation (Pratt) and observation at Kaliána is opposed however to the assumption, that the Himalaya Mountains are uncompensated.	43-44
April, 1901	(14)	The errors in Pratt's values of oceanic depths necessitate a re-calculation of the effects of the Himalayan mass and of the oceanic deficiency.	49
	(15)	A re-calculation shows that the Kaliána discrepancy is mythical.	90
October, 1901	(16)	The results of latitude observations at 160 stations do not justify any theory of entire compensation of mountains and seas; they support the suggestion made above in (12).	93
November, 1901	(17)	The results of two arcs of longitude however, observed across India from coast to coast, cannot be explained except on the hypothesis, that the Indian Ocean is wholly compensated.	96-98
	(18)	These longitude results, and the consequent adoption of the hypothesis, that the Indian Ocean is wholly compensated, necessitate the abandonment of the idea, that the change in the sign of the deflections along the parallel of 24° is due to the joint action of Himalayas and Ocean.	100
	(19)	If the Ocean is wholly compensated, and the Himalaya Mountains uncompensated, the latter will cause negative deflections north of latitude 24° and positive south of latitude 24°: observations however indicate the opposite, viz., positive north of 24°, negative south of 24°*.	103
	(20)	The evidence is conflicting: the large deflections in Sub-Himalayan regions preclude the hypothesis, that the Himalayas are wholly compensated: the great preponderance of negative deflections south of 24° seems to indicate entire Himalayan compensation.	100-103
	(21)	It is suggested that the change in the sign of the deflections along the parallel of 24° is not real, and is due to errors of the ellipsoid of reference.	104

* Kaliánpur being the station of reference, a "negative" deflection denotes a deflection more northerly than the deflection at Kaliánpur, and a "positive" deflection denotes a deflection more southerly than the deflection at Kaliánpur.

PREFACE.

Though the several parts of this paper constitute successive steps in one investigation, these steps were taken, so to speak, in the dark: one step led to another, and the earlier portions were written, before the calculations of Part IV had been begun. If the paper were to be rewritten, I should avoid in Part I an expression of views, which are subsequently shown to be untenable.

The problem of Himalayan attraction, after lying dormant for many years, has again forced itself on our attention: its solution requires more data than we yet possess, but a periodical investigation of the evidence is essential, if we wish to design the most profitable programmes of future work. Archdeacon Pratt considered that the attraction of the Himalayas was completely compensated, and that all discrepancies between astronomical and geodetic results might be attributed to *local* attractions. The examination of recent evidence suggests serious objections to the acceptance of this view. That the attraction of the Himalayas is partially compensated there are grounds for believing, but that their influence ceases within a hundred miles of their foot is a conclusion, that rests on no solid basis. I see nothing in the evidence to justify the belief, that Himalayan attraction is inappreciable at Cape Comorin.

A sense of loyalty to General Walker renders criticism of his theories an unwelcome task: such criticism can only be based on data, that were never at his disposal and that have been accumulated since his death. In 1896, a few months before his death, a paper by General Walker, containing a masterly summary of Indian geodetic work, was issued by the Royal Society: in this paper General Walker advocated the adoption of "groups" of astronomical stations for the purpose of eliminating the effects of "local attraction": he also in it explained certain perplexing phenomena by *assuming* the existence of a *southerly deflection* at Kaliánpur, *the station of reference of the Indian Survey*. In 1898 in consequence of this paper and in full accord with General Walker's views, we threw a "group" of astronomical stations round Kaliánpur: their results showed that the deflection at Kaliánpur due to local attraction was *northerly*: this unexpected issue created a dilemma: either General Walker had been mistaken in advocating "groups", or his *assumption* of a *southerly* deflection at Kaliánpur had been incorrect.

I have to acknowledge the great assistance which I have received from Mr. C. H. Mc A'Fee, Extra Deputy Superintendent, who was acting in charge of the Computing Office and who has superintended the publication of this pamphlet: the heights of compartments were averaged by Mr. J. Hickie, Extra Assistant Superintendent, now Chief Draftsman in Calcutta, and by Munshi Aulad Husain, Sub-Assistant Superintendent, and were compared with values obtained independently by myself: the computations have been carried out by Babu Shoshee Bhushan Shome, Senior Computer, and the effects of modifications in the ellipsoid of reference have been calculated by Babu Shiv Nath Saha, Second Computer.

By the kind permission of Colonel St. G. C. Gore, R.E., Surveyor General of India, I am enabled to attach as a frontispiece to this paper the cross-section of the outer Himalayas, which he has constructed from contoured maps.

I am much indebted to Mr. C. L. Griesbach, C.I.E., Director of the Geological Survey of India, for valuable advice and information.

DEHRA DÚN:
December 1901.

S. G. BURRARD.

CHART No. 3

Chart showing resultant Local attractions at the stations of the Kaliánpur Group, obtained from a combination of the results of the two preceding charts.

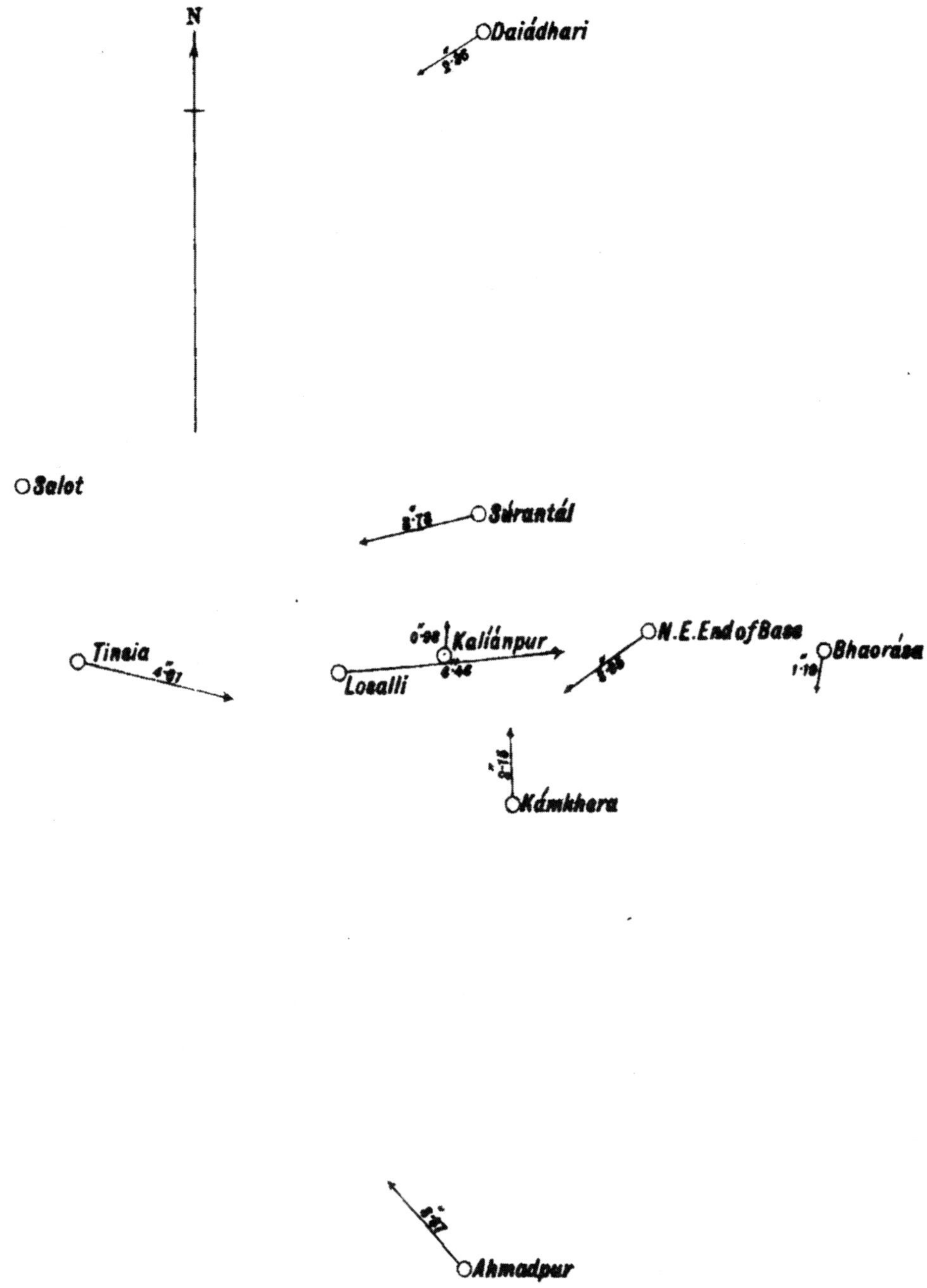

Scale of Plan 1 Inch = 12 Miles.

The arrows show the directions of the deflections of the Plumb-lines; the lengths of the arrows are proportional to the amounts of the deflections.

CHART No. 5

Deflections of the Plumb-line in the Prime Vertical.

DIAGRAM No. 3

The deflection at Kaliánpur derived from results of the Group.

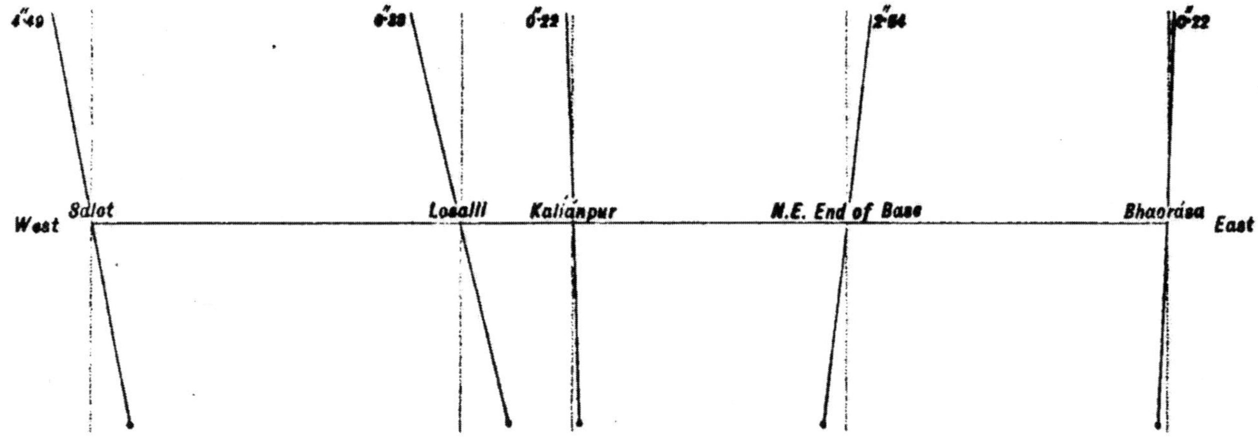

DIAGRAM No. 4

The deflection at Kaliánpur derived from results of all India.

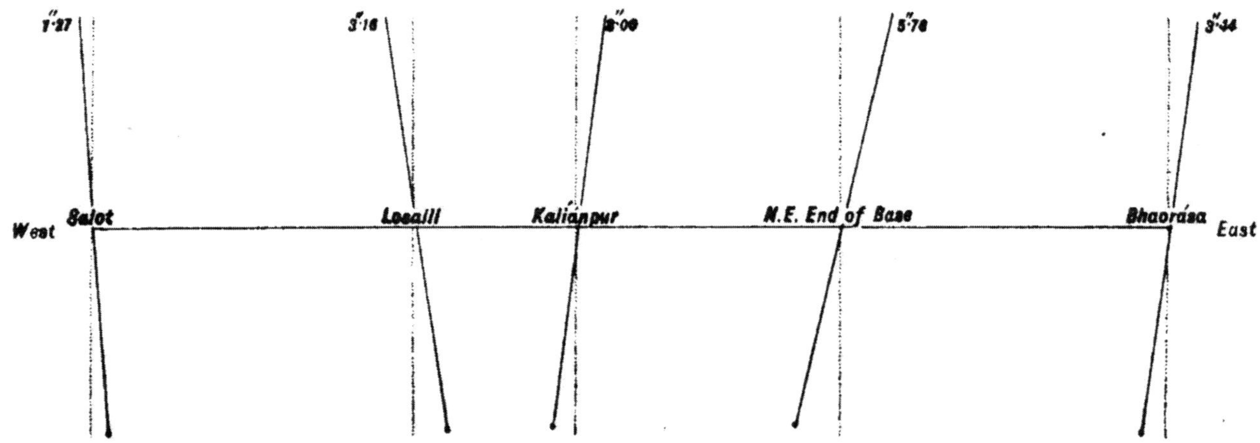

Photozincographed at the Office of the Trigonometrical Branch, Survey of India, Dehra Dún, July, 1903.

No. D. 6, S. I. 10 - 2 - 1903 - 100

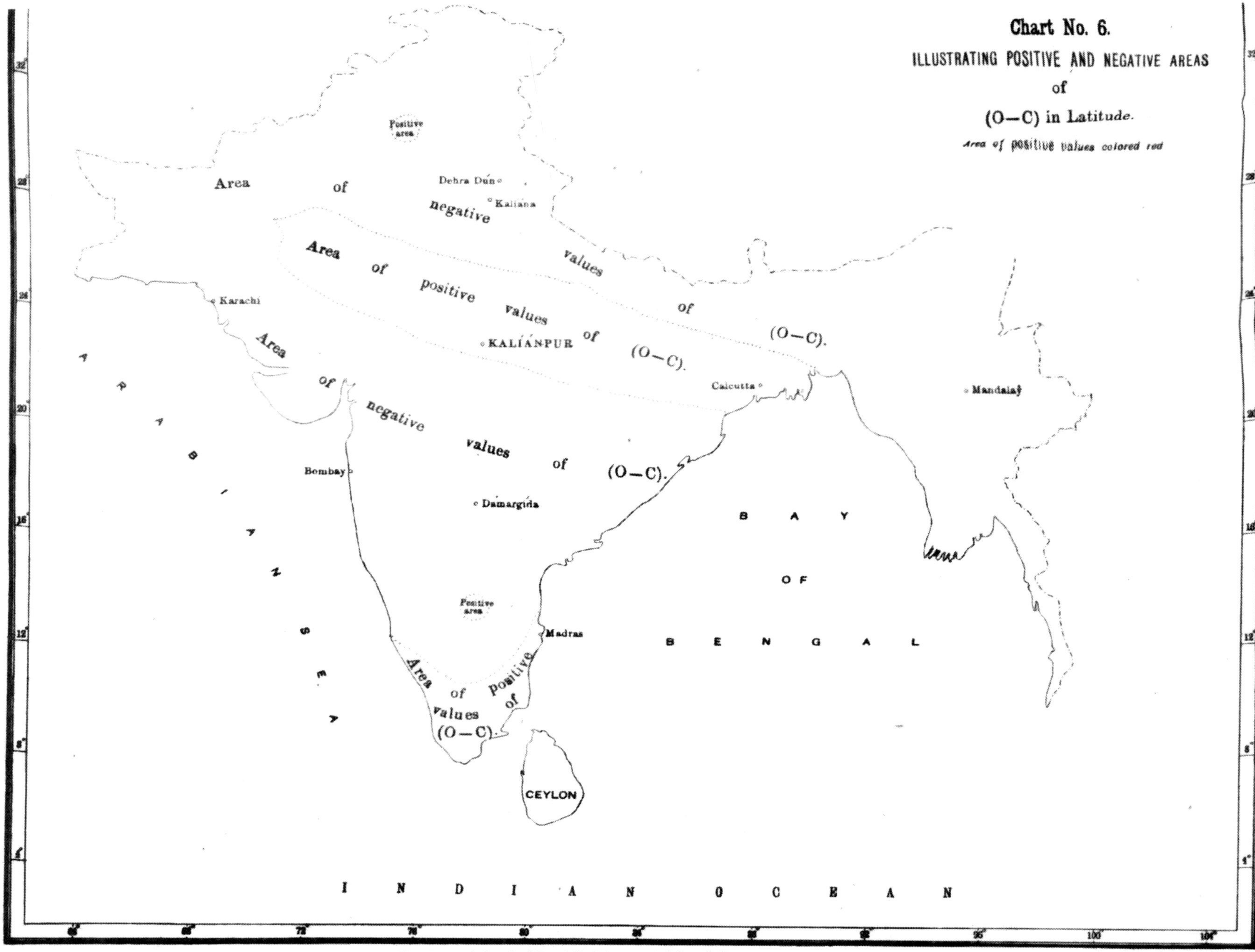

Regr. No. D. 297, S. I. D.-April 1901 – 150
" 7 " -Feby 1902 – 100

Photozincographed at the Office of the Trigonometrical Branch, Survey of India, Dehra Dún.
May 1901.

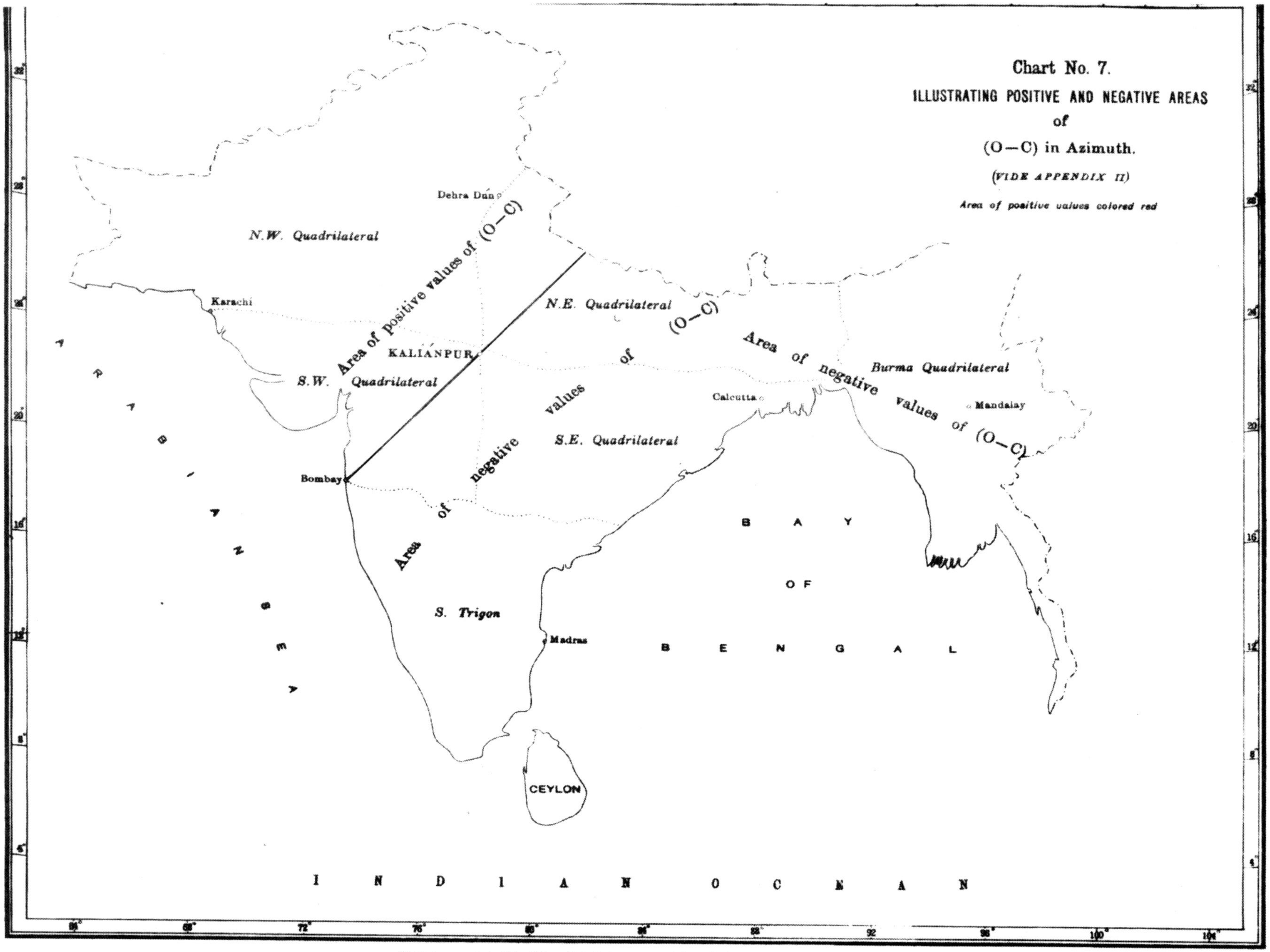

Photozincographed at the Office of the Trigonometrical Branch, Survey of India, Dehra Dún.
May 1901.

Regr No. D. 298. S. I. D. - April 1901 - 150
" 8 " - Feby. 1902 - 100

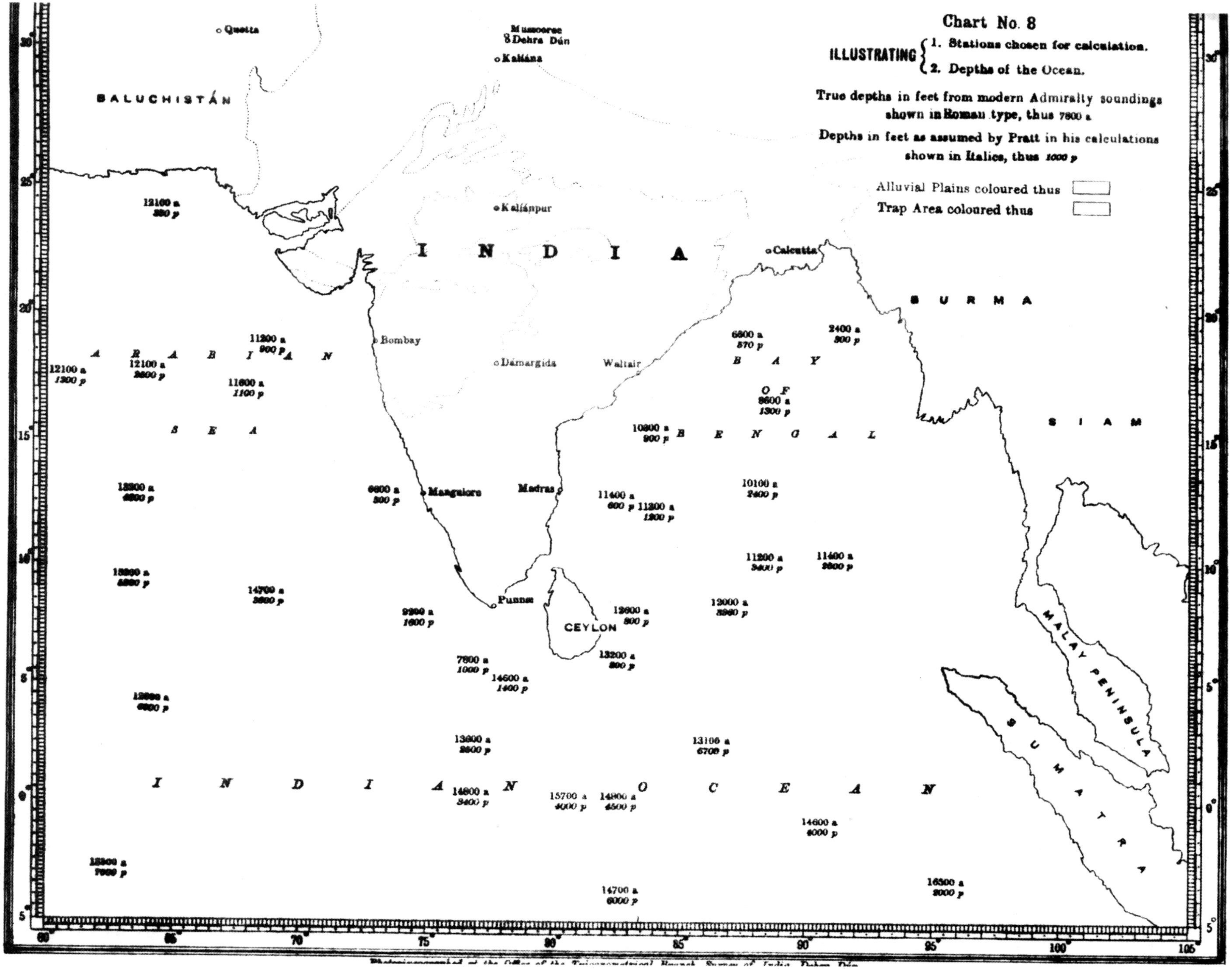
Chart No. 8
ILLUSTRATING 1. Stations chosen for calculation.
2. Depths of the Ocean.
True depths in feet from modern Admiralty soundings shown in Roman type, thus 7800 a
Depths in feet as assumed by Pratt in his calculations shown in Italics, thus 1000 p
Alluvial Plains coloured thus
Trap Area coloured thus
BALUCHISTÁN
Quetta
Mussooree
Dehra Dún
Kaliána
Kaliánpur
INDIA
Calcutta
BURMA
SIAM
Bombay
Dámargida
Waltair
Mangalore
Madras
Punnæ
CEYLON
MALAY PENINSULA
SUMATRA
ARABIAN SEA
BAY OF BENGAL
INDIAN OCEAN

Chart No. 9

ILLUSTRATING

THE SITUATION OF DEHRA DUN WITH REGARD TO THE TIBETAN PLATEAU

A cross-section of the Himalayas on the Great Circle through Dehra Dun in Azimuth 30° East of North, as constructed from the maps of the Indian Survey by Colonel St. G. C. Gore, R.E., Surveyor General of India.

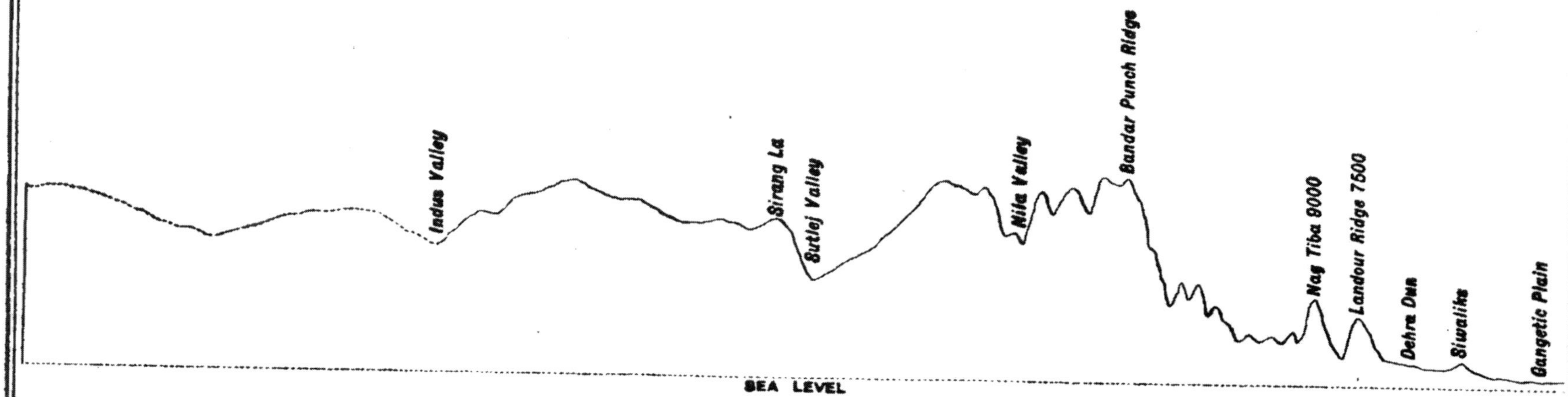

SCALES

Horizontal 1 Inch = 32 Miles

Vertical 1 Inch = 16000 Feet

Regr. No. D 540, S. I. D.- July 1901 - 750
" 10 " - Feby. 1902 - 100

Photozincographed at the Office of the Trigonometrical Branch, Survey of India, Dehra Dún, August 1901.

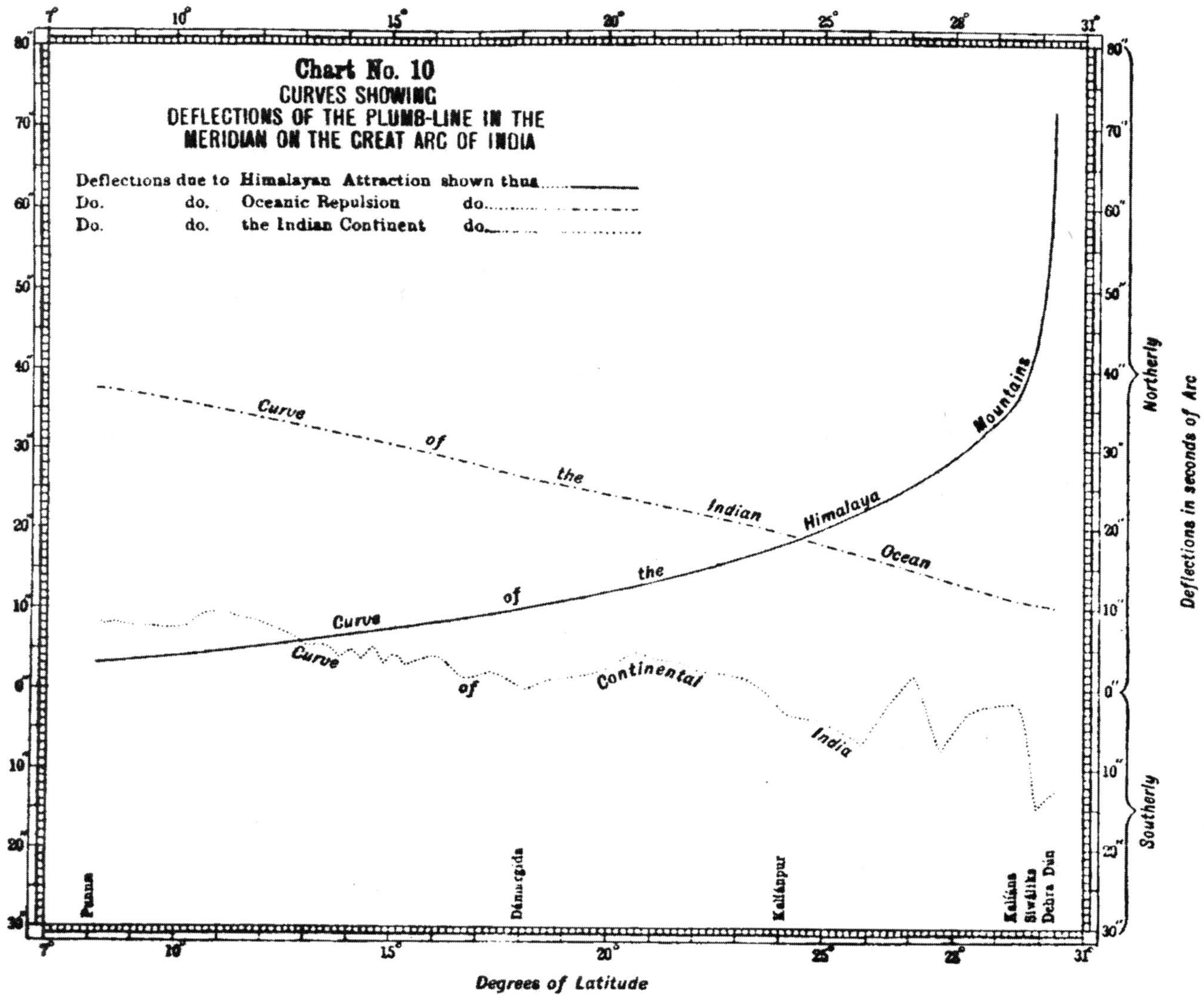

Photozincographed at the Office of the Trigonometrical Branch, Survey of India, Dehra Dún.
August 1901.

Regr. No D. 542, S. I. D. - July 1901 – 150

" " " - Feby. 1902 – 100

SECTION ACROSS INDIA

ON THE PARALLEL OF 31°

Showing apparent deflections of the plumb-line

at

AMRITSAR and MOOLTAN

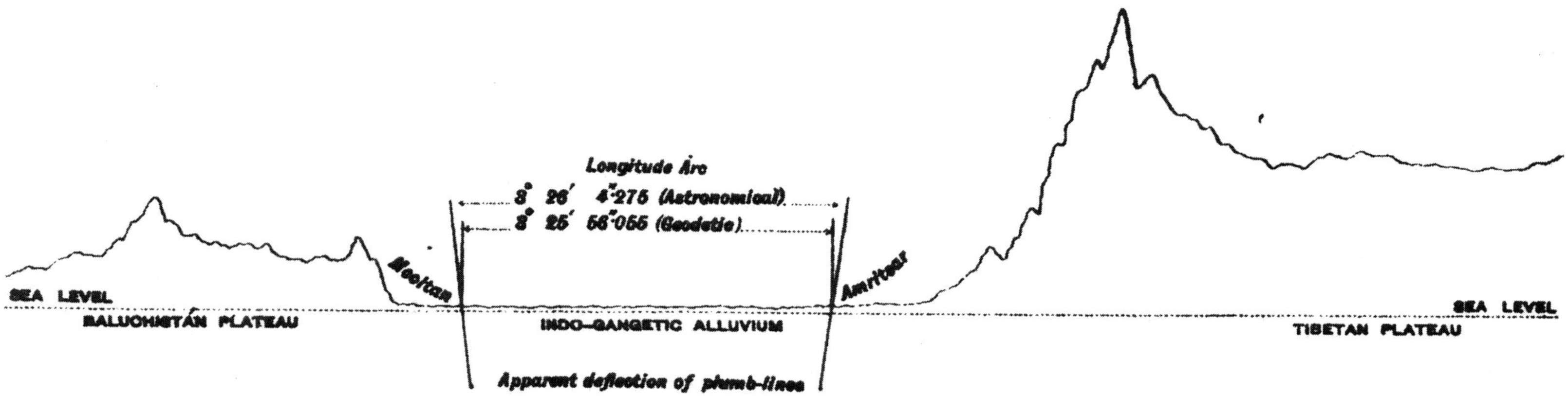

SECTION ACROSS INDIA

ON THE PARALLEL OF 13°

Showing apparent deflections of the plumb-line

at

MADRAS and MANGALORE

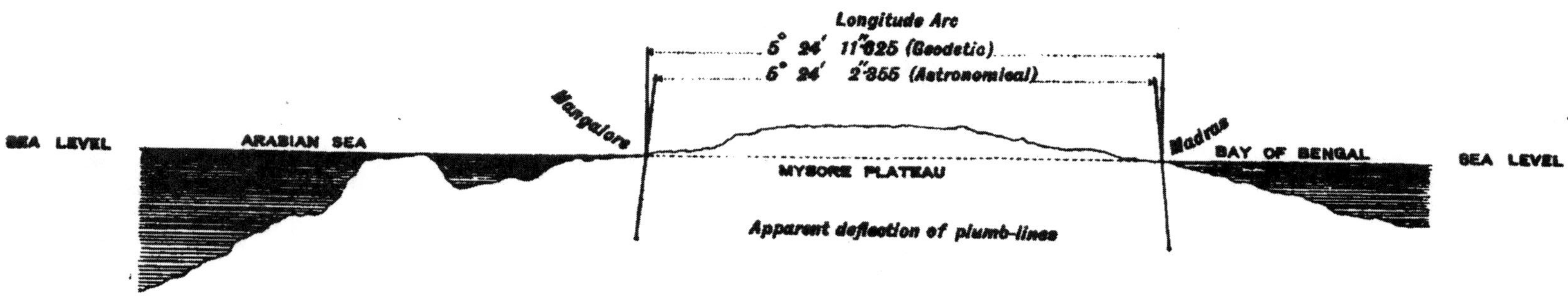

SCALES

Horizontal	Vertical
1 Inch = 128 Miles	1 Inch = 16000 Feet

Photozincographed at the Office of the Trigonometrical Branch, Survey of India, Dehra Dún.
July 1901.

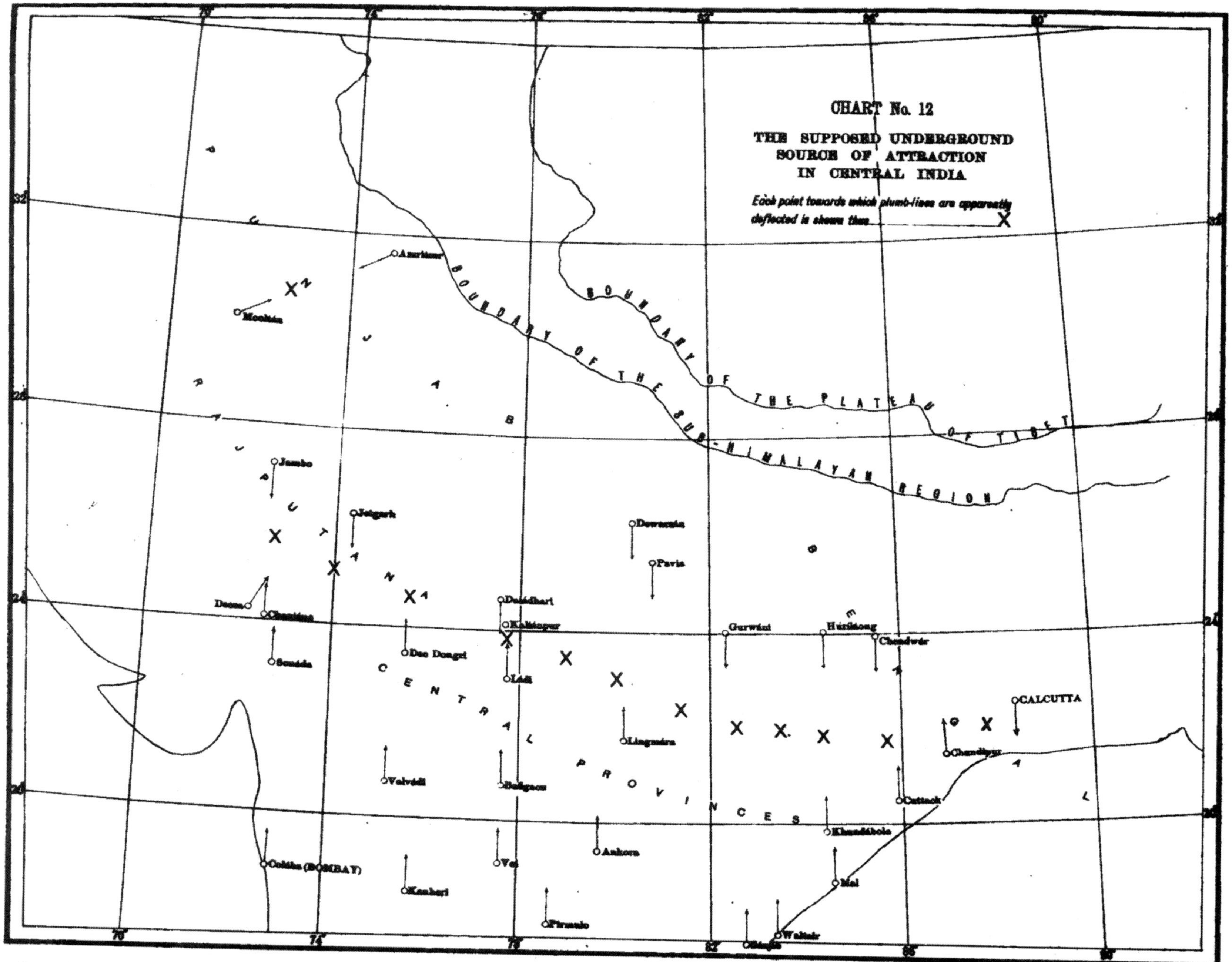

No. D. 13, S. I. 10 — 3 — 1902 — 100

Photozincographed at the Office of the Trigonometrical Branch, Survey of India, Dehra Dûn, January 1904.

PART I.

On the errors of the initial values of Latitude and Azimuth in India.

In 1898-99 a group of latitudes and azimuths was observed round Kalíánpur by Captain Lenox Conyngham. These observations are very important, and it is desirable to review the results obtained for the purpose of discovering the most profitable directions for future progress.

Groups of latitudes round a central station were first observed in India by Colonel John Herschel, R.E., in 1870, and are considered to give a more reliable value of the local attraction at the central station than observations taken at the central station itself. On page 807 of *India's Contributions to Geodesy*, General Walker writes "Before mathematical treatment can be "advantageously commenced, steps should be taken to diminish the local deflections by which the "observations are burdened. In treating meridional arcs, the only possible way is to combine a "number of the astronomical stations within a narrow belt of parallel together, and take the "mean latitude of the group: this gives us combinations of data which will be far more valuable "for mathematical treatment than the separate individual initial data, because the mean astrono-"mical latitude of a number of points may certainly be assumed to be far more free from deflec-"tion than the latitude of any single point".

On page [155] of the Geodetic Survey of South Africa 1896, Sir David Gill argues in favour of groups: he recommends that Principal Stations should be surrounded by six astronomical stations; "If the astronomical latitudes and longitudes of these points", he writes, "are "then determined, we have, from a discussion of the discordances between the geodetic and "astronomical results of the figure, all the requisite data for computing the local attraction at the "central point". On page [157] Sir David Gill foresees, that if local attraction is persistent in one direction over large continuous areas, group observations would not avail to eliminate its effects, and he recommends in such cases, that the group be extended, till the centre of maximum deviation has been located and surrounded with astronomical points.

The Local Attraction at Kalianpur deduced from the observations of all India.

The local attraction at Kaliánpur was deduced by General Walker, before the recent surrounding group of latitudes and azimuths had been observed. It had always been noticeable, that the value of (O – C)* in latitudes had a tendency to be negative, and General Walker explained this tendency by the theory, that the plumb-line at Kaliánpur was deflected 2″ to the south: on page 804 of *India's Contributions to Geodesy* he writes "Of the 148 astronomical latitudes available "there are 90 cases of negative excess to 58 of positive excess. If the latitude of Kaliánpur is "diminished by 2″·0, the whole of the geodetic latitudes will be correspondingly diminished, and "this will make the number of positive and negative cases almost exactly equal †".

In the case of the fundamental *latitude* General Walker estimated the deflection of the plumb-line *after* the Indian triangulation had been computed, but the effect of local attraction on the fundamental *azimuth* he deduced, *before* the reduction of the triangulation had been carried out. Thus the triangulation is based on an observed latitude, uncorrected for local attraction, and on a derived azimuth, corrected for local attraction. The computed geodetic values of latitude are based on the observed latitude of Kaliánpur, but the computed geodetic values of azimuth are based on the derived azimuth of Kaliánpur. On pages 137 to 141 of Volume II of the Great Trigonometrical Survey of India, General Walker explains, that of 35 stations situated in different parts of India, at which astronomical azimuths had been observed, the value of (O – C) proved to be negative at 26, and positive at 9, and he arrives at the conclusion—a conclusion generally accepted hitherto—that the adopted value of the fundamental azimuth at Kaliánpur was too great. He then collected all the stations at which azimuths had been observed, omitting those near the Himalayas, and finally derived a correction of—1″·1 to be applied to the observed azimuth at Kaliánpur, thus arguing a local attraction of $1''\cdot1 \cot \phi = 2''\cdot45$ to the west.

On page 446 of Volume XV, Great Trigonometrical Survey of India, Colonel Strahan, following General Walker, deduces the probable deflection of the plumb-line in the Prime Vertical at Kaliánpur from a comparison of the 55 astronomical Arcs of Longitude, measured over India and Burma, with their geodetic values: he finds the zenith at Kaliánpur to be probably displaced 3″·42 to the eastwards, or in other words a local attraction of the plumb-line of about 3″ to the westwards.

---

* In determining deflections of the plumb-line we deduct the geodetic value from the astronomic: the astronomic value is *observed*, and designated O: the geodetic value is computed through the triangulation from the initial elements and designated C: the deflection of the plumb-line at any station is taken to be (O – C), the assumption being made that C or the geodetic value is the true one. In the case of latitudes, if the plumb-line at any station is attracted to the north, the zenith will be displaced to the south, the observed latitude or O will be too small, and (O – C) will be negative.

In the case of azimuths, if the plumb-line at any station is attracted to the east, the zenith will be displaced to the west, the observed azimuth or O measured from south by west will be too small, and (O – C) will be negative.

We therefore have the following rules of signs, always assuming that the value of C is correct:—

If at any station (O – C) in latitude is negative, the attraction in the meridian is northerly
If (O – C) in latitude is positive, the attraction in the meridian is southerly
If (O – C) in azimuth is negative, the attraction in the Prime Vertical is easterly
If (O – C) in azimuth is positive, the attraction in the Prime Vertical is westerly.

† In comparing the number of cases of negative and positive excess it is difficult to decide whether to reject certain latitudes or not: it frequently happens that two or more astronomical latitudes have been observed within a few miles of each other, and it is questionable in discussing the latitude of all India whether we should regard each value as an independent latitude, or whether we should adopt the mean of the group. Two latitudes were observed within 5 miles of one another at Madras (*see* page 782 of *India's Contributions to Geodesy*): six latitudes were observed within a radius of 4 miles near Punnæ (*see* group 1, page 778); the four latitudes of group 2, and the three latitudes of group 3, page 778, are other cases in point: when General Walker gives 90 and 58 as the respective numbers of negative and positive excess, he is giving full weight to every observed value of every group; if we take the mean of a group spread over a small area as a single value of latitude, and thus give equal weights to equal areas, we find that there are, including the latitudes observed in the last two years, 111 instances of negative excess and 49 instances of positive excess. Though this difficulty in the matter of combinations and rejections does exist, General Walker's deduction of the meridional deflection at Kaliánpur from the Indian latitudes as a whole has met with general approval, and of recent years a southerly attraction of 2″ at Kaliánpur has been accepted as a working hypothesis in explanation of the excess of negative values.

Before the group of latitudes and azimuths had been observed round Kaliánpur, the local attraction at Kaliánpur had thus been deduced by Indian geodesists from the results of 148 observed latitudes, of 51 observed azimuths, and of 55 arcs of longitude, distributed over India with the following results :—

Deflection of the plumb-line in the meridian = 2" to the south

Deflection of the plumb-line in the Prime Vertical (as deduced from Azimuths) = 2½" to the west

Deflection of the plumb-line in the Prime Vertical (deduced from Longitudes) = 3" to the west.

It remains now to be seen to what extent these results have been corroborated by Captain Lenox Conyngham's group round Kaliánpur.

The Observed Latitude of Kalianpur.

The Kaliánpur observations can best be analysed in the following order :—

(a) The latitude at Kaliánpur derived from observations at Kaliánpur only.

(b) The latitude at Kaliánpur derived from observations at the group of surrounding stations.

(c) The azimuth at Kaliánpur derived from observations at Kaliánpur only.

(d) The azimuth at Kaliánpur derived from observations at the group of surrounding stations.

The latitude of Kaliánpur itself has now been observed on six occasions as follows* :—

Date	Observer	Value
1824-25 ...	Geo. Everest ...	24° 7′ 10″·76 ± 0″·13
1839-40 ...	Andrew Waugh ...	10 ·92 ± 0 ·08
1840-41 ...	Geo. Everest and T. Renny-Tailyour ...	11 ·18 ± 0 ·07
February 1865 ...	W. M. Campbell ...	11 ·44 ± 0 ·07
November 1865 ...	W. M. Campbell ...	10 ·90 ± 0 ·07
1898-99 ...	G. Lenox Conyngham	10 ·59 ± 0 ·08
Mean ...	...	24° 7′ 10″·97

* Lenox Conyngham's value is corrected for variation of latitude from the data furnished in Albrecht's Report, Potsdam, February 1900: the earlier values are uncorrected.

The adopted initial value of latitude for the Indian Survey is 24° 7′ 11″·26: this value was derived by Everest from the results of 1824, '25, '39, '40 and '41, using the best values of the stars' places, that he could then obtain: his stars' places have since been revised, and Everest's mean value of latitude, though still the fundamental latitude of India, is no longer deducible from the observations. The initial latitude of India therefore appears to be 0″·29 too great, owing to errors of observation and star's place.

The Group of Latitudes round Kalianpur.

Astronomical latitudes have been recently observed by Captain Lenox Conyngham at 8 stations round Kaliánpur (*see* Chart No. 1).

The results are as follows :—

Station	Observed Latitude	Correction to Mean Pole	Seconds of Corrected Observed Latitude = O	Seconds of Computed Geodetic Latitude = C	O − C = Deflection of Plumb-line in Meridian
	° ′ ″	″	″	″	″
Daiádhari ...	24 38 18·83	+ 0·06	18·89	17·57	+ 1·32
Súrantál ...	14 21·41	+ 0·14	21·55	20·42	+ 1·13
Sironj, N. E. End Base	8 55·46	+ 0·11	55·57	53·57	+ 2·00
Bhaorása ...	8 5·13	+ 0·08	5·21	3·73	+ 1·48
Losalli ...	6 18·31	+ 0·15	18·46	19·17	− 0·71
Tinsia ...	6 29·11	+ 0·15	29·26	27·97	+ 1·29
Kámkhera ...	23 59 42·95	+ 0·14	43·09	44·93	− 1·84
Ahmadpur ...	36 18·59	+ 0·11	18·70	20·88	− 2·18

The geodetic latitudes in this table have been computed on the assumption, that the latitude of Kaliánpur is 24° 7′ 11″·26. It is significant that the positive values of (O − C) should be in excess of the negative, the results of the group being thus in opposition to those of all India.

In the following table the latitude of Kaliánpur is deduced from the observed latitude at each station, by applying the geodetic difference of latitude derived from the triangulation :—

Station of Observation	Observed Latitude	Geodetic Difference	Resulting Latitude of Kaliánpur
	° ′ ″	′ ″	° ′ ″
Daiádhari ...	24 38 18·89	−31 6·31	24 7 12·58
Súrantál ..	14 21·55	− 7 9·16	12·39
Sironj, N.E. End Base	8 55·57	− 1 42·31	13·26
Bhaorása ...	8 5·21	− 0 52·47	12·74
Losalli ...	6 18·46	+ 0 52·09	10·55
Tinsia ...	6 29·26	+ 0 43·29	12·55
Kámkhera ...	23 59 43·09	+ 7 26·33	9·42
Ahmadpur ...	36 18·70	+30 50·38	9·08
Mean of Lenox Conyngham's group ...			24° 7′ 11″·57

We have now the three following values of the latitude of Kalíánpur :—

Value adopted in computations of the triangulation	**24° 7′ 11″·26**
Mean observed value	**24 7 10 ·97**
Value derived from the group	**24 7 11 ·57**

On the assumption that the value derived from the group is freed from the effect of local attraction, we deduce that the astronomical zenith at Kalíánpur is displaced 0″·60 to the south, and that *there is a deflection of the plumb-line in the meridian at Kalíánpur of 0″·60 to the north.*

It has been explained, that General Walker, arguing from all India, estimated the meridional deflection of the plumb-line at Kalíánpur at 2″ to the south. If we correct the initial latitude at Kalíánpur, firstly, by −0″·29 for error for star's place and observation, and, secondly, by +0″·60 for local attraction as derived from the group, and introduce the value 24° 7′ 11″·57 into the computations, the excess of negative values of (O − C) in India over positive is increased; there will be then 117 cases of a negative value, and 43 cases of a positive value, and *the mean magnitude of the negative values will be considerably greater than the mean magnitude of the positive values.*

The Fundamental Azimuth.

The azimuth of Súrantál has been observed from Kalíánpur on two occasions with the following results* :—

Date	Observer	Value
		° ′ ″
1836	Geo. Everest ...	190 27 6·20
1898-99	G. Lenox Conyngham	190 27 6·37
	Mean	190° 27′ 6″·29

General Walker's value of the fundamental Azimuth, derived from azimuths observed in different parts of India, is 190° 27′ 5″·10 or 1″·19 less than the latest mean observed value.

The Group of Azimuths round Kalianpur.

Astronomical Azimuths have been recently observed by Captain Lenox Conyngham at 8 stations round Kalíánpur.

* Lenox Conyngham's value is corrected for variation of latitude from the data furnished in Albrecht's Report, Potsdam, February 1900.

The results are as follows :—

Station	Observed Azimuth	Correction to Mean Pole	Seconds of Observed Corrected Azimuth = O	Seconds of Computed Geodetic Azimuth = C	(O − C)
	° ′ ″	″	″	″	″
Daiádhari ...	303 32 52·53	+ 0·15	52·68	50·41	+ 2·27
Súrantál ...	10 27 43·37	+ 0·02	43·39	40·46	+ 2·93
Sironj, N. E. End Base	80 46 33·96	+ 0·08	34·04	31·61	+ 2·43
Bhaorása ...	95 12 39·36	+ 0·11	39·47	38·08	+ 1·39
Losalli ...	305 52 55·80	− 0·07	55·73	57·30	− 1·57
Salot ...	175 58 10·16	...	10·16	10·89	− 0·73
Kámkhera ...	154 45 36·67	− 0·05	36·62	35·31	+ 1·31
Ahmadpur ...	185 10 56·27	− 0·08	56·19	53·91	+ 2·28

The Geodetic Azimuths have been computed on the assumption that the azimuth of Súrantál at Kalíánpur is 190° 27′ 5″·10. Again the positive values of (O−C) exceed the negative, in opposition to the experience of all India.

In the following table the value of the fundamental azimuth at Kalíánpur is deduced from the observed azimuth at each station of the group by applying the geodetic difference of azimuth derived from the triangulation.

Station of Observation	Observed Azimuth	Geodetic difference	Resulting Azimuth at Kalíánpur
	° ′ ″	° ′ ″	° ′ ″
Daiádhari ...	303 32 52·68	113 5 45·31	190 27 7·37
Súrantál ...	10 27 43·39	179 59 24·64	8·03
Sironj, N.E. End Base	80 46 34·04	109 40 33·49	7·53
Bhaorása ...	95 12 39·47	95 14 27·02	6·49
Losalli ...	305 52 55·73	115 25 52·20	3·53
Salot ...	175 58 10·16	14 28 54·21	4·37
Kámkhera ...	154 45 36·62	35 41 29·79	6·41
Ahmadpur ...	185 10 56·19	5 16 11·19	7·38
	Mean ...	...	... 190° 27′ 6″·39

We have now the three following values of the azimuth at Kalíánpur:—

Value adopted in computations of the triangulation 190° 27′ 5″·10

Mean observed value ... ... 6·29

Value derived from the group ... ... 6·39

On the assumption that the value derived from the group is freed from the effect of local attraction, we deduce that the astronomical zenith at Kalíánpur is displaced $0''\cdot 10 \times \cot \phi = 0''\cdot 22$ to the west, that there is a deflection of the plumb-line in the Prime Vertical at Kalíánpur of 0″·22 to the east, that the fundamental azimuth is 1″·29 too small, and that every value of (O − C) requires a correction of − 1″·29. A complete list of the Observed Azimuths is published as Appendix II of this paper. It will be seen from that list, that if Walker's value of the initial azimuth be adhered to, there are 150 negative values and 59 positive values of (O − C)*, and that if the mean value derived from the group be accepted, *there will be 170 negative values and 39 positive values.*

Deflection of the Plumb-line at Kalianpur.

If the mean latitude and azimuth, obtained from the group, be assumed freed from the effect of local attraction, the deflection of the plumb-line at the several stations of the group may be stated as follows:—

Station	Deflection of Plumb-line		
	Effect on Azimuth $= (O_A - C_A)$	Deflection in Prime Vertical $= (O_A - C_A)\cot\phi$	Deflection in Meridian $= (O_\phi - C_\phi)$
	″	″	″
Daiádhari ...	+ 0·98	+ 2·13 W.	+ 1·01 S.
Súrantál ...	+ 1·64	+ 3·64 W.	+ 0·82 S.
N. E. End of Base ...	+ 1·14	+ 2·54 W.	+ 1·69 S.
Bhaorása ...	+ 0·10	+ 0·22 W.	+ 1·17 S.
Kalíánpur ...	− 0·10	− 0·22 E.	− 0·60 N.
Losalli ...	− 2·86	− 6·38 E.	− 1·02 N.
Tinsia ...	...	...	+ 0·98 S.
Salot ...	− 2·02	− 4·49 E.	...
Kámkhera ...	+ 0·02	+ 0·04 W.	− 2·15 N.
Ahmadpur ...	+ 0·99	+ 2·27 W.	− 2·49 N.

* On page 804 of India's Contributions to Geodesy, General Walker explains, that this predominance of negative values can be eliminated by increasing the negative correction applied to the initial azimuth. But according to the evidence of the group the correction to the initial azimuth should be positive.

The deflection of the plumb-line at Kaliánpur itself has thus been given as follows:—

	In the Meridian	In Azimuth	In the Prime Vertical
By the group of latitudes round Kaliánpur	" 0·60 North		
By the latitudes of all India ...	2·00 South		
By the group of azimuths round Kaliánpur		" 0·10 East	" 0·22 East
By the azimuths of all India ...		1·19 West	2·65 West
By the longitudes of all India ...			3·00 West
Difference	2″·60	1″·29	2″·87

In each case the results of the group have falsified predictions based with confidence on the results of all India*.

The contradictions with which we are now faced, cause us to consider the following questions:—

(a) Can any inequality in the distribution of matter in the immediate locality of Kaliánpur be conceived, that can deflect the plumb-line 3″·5 to the south-west† at Kaliánpur itself, and yet allow the mean deflection, resulting from eight surrounding stations in the vicinity, to be half a second‡ to the north-east?

(b) Can there exist any external source of attraction affecting the plumb-line at every station of the group, and rendering the mean determination of the deflection in error by 4″?

Explanation of Charts 4 and 5.

In Charts 4 and 5 are shown four diagrams:—

The first diagram gives a vertical section through stations on the meridian of Kaliánpur.§ The deflections of the plumb-line in the meridian on an exaggerated scale are shown in this diagram on the assumption, that the mean of the group of latitudes is freed from the effect of local attraction.

All the plumb-lines tend inwards: if their positions are to be explained on the hypothesis of some local irregularity of matter, the assumption, that appears least objectionable, seems to be, that Kaliánpur is situated over the centre of a subterranean mass of excessive density attracting the plumb-line at Ahmadpur and Daiádhari inwards. *This assumption would not account for the excess of negative values in India.*

If the positions of the plumb-lines in Diagram No. 1 are to be explained on the hypothesis of some external force, it seems necessary to assume, that a source of repulsion exists at

---

* It is worth noting that the deflection in azimuth × cotangent latitude = deflection in prime vertical, and that 1″·19 × cot 24° 7′ = 2″·65, and that therefore the deflection in the Prime Vertical derived from the azimuths of India is 2″·65 west, or only 0″·35 less than the deflection of 3″ west derived independently from the longitudes of India.

† That is 2″ south on the Meridian, and 3″ west on the Prime Vertical.

‡ 0″·60 north on the Meridian and 0″·22 east on the Prime Vertical.

§ A description of Kaliánpur and the surrounding country is given in Appendix I.

a considerable distance either to the north or south of the group. Such a source might affect the mean value of the group by 10″ or more, but not be shown up in the results. If it existed to the south, the plumb-line at Ahmadpur, the nearest station would be repelled slightly more than that at Kaliánpur, and that at Daiádhari slightly less. The astronomical latitudes observed at stations of the Great Arc immediately south of Ahmadpur afford no evidence of the existence of any such source of repulsion, *vide* Table of Latitudes, which follows page 14.

We can redraw the first diagram on the supposition that a southerly attraction of 2″ exists at Kaliánpur, as was deduced by General Walker from a consideration of the Indian latitudes as a whole: this has been done in Diagram No. 2.

A source of attraction south of Ahmadpur would deflect the nearer plumb-lines more than the further: a source of repulsion north of Daiádhari might produce the deflections in this diagram.

The existence of a source of repulsion north of Daiádhari is not confirmed by the azimuth observations at Salot, but is not incompatible with the results of the latitudes observed at Kesri and Usira, *vide* Table following page 14. Possibly a further group of latitudes may locate such a source. The proof of its existence will be our justification for the retention of Walker's correction of − 2″ to geodetic latitudes.

The third diagram gives a vertical section through stations on the Prime Vertical of Kaliánpur: the deflections of the plumb-lines in the Prime Vertical on an exaggerated scale are shown on the assumption, that the mean of the group of azimuths is freed from the effect of local attraction.

All the plumb-lines again tend inwards. *This diagram will not explain the excess of negative values of (O − C) in azimuth,* that prevails throughout India and Burma.

We can redraw the third diagram on the supposition that a westerly attraction of 3″ exists at Kaliánpur, as has been deduced by Walker and Strahan from a consideration of the Indian azimuths and longitudes: this has been done in Diagram No. 4.

This diagram would show that a source of attraction exists between Kaliánpur and Losalli.

The Calculations of Archdeacon Pratt.

Before, however, we endeavour to decide whether the contradictions at Kaliánpur are due to local or external causes, it will be well to consider the external forces, that affect the plumb-line in India, and as an indispensable aid to this investigation, to recall the calculations and theories of Archdeacon Pratt*.

In 1852 the Ven. John Henry Pratt, Archdeacon of Calcutta, was asked by Sir Andrew Waugh, Surveyor General of India, to turn his attention to the influence of Mountain attraction upon the operations of the Great Trigonometrical Survey of India. It had been pointed out by Everest in his Great Arc of India, 1847, that if the curvature of the Indian Arc be taken the same as that of the mean figure, the observed latitude of Kaliána, a station on the Great Arc 5¼ degrees north of Kaliánpur, was 5″·236 less than its geodetic latitude, and the observed latitude of

* Of recent years we have been possibly too apt to attribute differences between astronomical and geodetic values to mere local deviations of gravity and to regard them as due to local irregularities in the distribution of matter in the immediate neighbourhood of the stations of observation. The method of treating these differences by minimum squares can only be justified on the ground that they are purely local and accidental, and its practice has tended to strengthen the belief in their irregularity. The discovery, moreover, that deflections of the plumb-line occur in flat unbroken plains and the theory, which it necessitated, that these deflections are due to invisible subterranean causes, have also helped of late years gradually to give rise to the idea, now generally prevalent, that local attractions obey no explicable law and that no result however contradictory need excite surprise. It is only in the presence of some enormous visible mass, such as the Himalayan Mountains, and when large constant deflections of gravity occur, that an external source of attraction affecting large areas is admitted, and that the method of minimum squares is considered inapplicable.

Dámargída, a station on the Great Arc, six degrees south of Kaliánpur, was 3″·791 less than its geodetic latitude. The problem, that Waugh set Pratt to solve, was to calculate by some direct method the actual amount of the attraction of the Himalayan mass, and of the deflection given by it to the plumb-line. The results are shown at page 85 of Pratt's first paper, Philosophical Transactions of the Royal Society, 1854, to be as follows:—

Deflection of the plumb-line in the Meridian
at Kaliána . . . . 27″·853 North.
at Kaliánpur . . . . 11″·968 North.
at Dámargída . . . . 6″·909 North.

Deflection of the plumb-line in the Prime Vertical
at Kaliána . . . . 16″·942 East.
at Kaliánpur . . . . 4″·763 East.
at Dámargída . . . . 2″·723 East.

Total deflections of the plumb-line
at Kaliána . . . . 32″·601 in azimuth 31° 18′ East of North
at Kaliánpur . . . . 12″·880 in azimuth 21° 42′ „ „
at Dámargída . . . . 7″·426 in azimuth 21° 31′ „ „

This calculation increased the difficulty, which it was intended to remove, as the disturbing effect of the Himalayas was shown to be greater in amount than had ever been anticipated.

It may be objected now that Pratt's knowledge of the mass and density of the Himalayas was deficient, and that a recalculation based on modern data might reduce Pratt's values of the deflections. Pratt took the *density* of the Himalayas at 2·75, whereas Mr. C. L. Griesbach, the present Director of the Geological Survey of India, informs me, that the mean density probably lies between 2·6 and 2·7: if we reduce the value of the density from 2·75 to 2·65 in Pratt's formulæ, his deflections will be reduced by only one-twenty-fifth part. In the matter of *area* Pratt's southern limits of the Himalayan Range are geographically correct; the accuracy of his northern limits of the Tibetan plateau and of his position of the Altai Mountains is not very material, as the distant ranges exercised but slight effect on his results: he omitted the Hindu Kush and the Sulemán Mountains, and he placed the Kuen Luen Range perhaps 100 miles too far north: an examination of Pratt's calculations teaches, that no reduction in the values of his deflections can be expected to result from the comparatively trifling corrections, which modern geographical knowledge might apply to his Himalayan and Tibetan *areas*.

In the matter of *heights* Pratt shows himself, that the deflections are more due to the table-lands and to the plateaus than to the higher and more prominent snow peaks: he takes the line joining Leh and Lhasa to be 10,000 feet high, and he assumes that the Tibetan plateau slopes gradually down to the north, and is 4,000 feet high in latitude 40°: the modern value of the height of Leh is 11,000 and of Lhasa 12,000 feet and the line joining them is known now to cross altitudes of 15,000 feet. Modern maps show heights of 15,000 and 14,000 feet in the centre of Tibet, where Pratt showed 7,000 only. No alteration of Pratt's *heights*, such as can be justified by modern explorations, will reduce his values of the deflections.

Pratt's paper was answered by Sir George Airy, who suggested that there was probably a deficiency of matter beneath the mountains, which counteracted their effect upon stations in the plains: in a postscript to a paper on the English Arc of Meridian (Philosophical Transactions of the Royal Society, 1855) Pratt states his objection to Airy's hypothesis, and gives his opinion, that the only explanation of the discrepancy, between his calculation and the results of the Indian Survey, is to be found in the curvature of the Indian Arc being greater than that of the Mean Figure.

In 1858 Pratt reduced to calculation (Philosophical Transactions of the Royal Society, 1858) another hypothesis regarding deficiency of matter below the mountains, *viz.*, that the irregularities of the mountain surface have arisen from the expansion upwards of the crust of the earth from depths below, which has upheaved the mountains and produced a slight but extensive attenuation of the mass below them: his calculations furnished the following results.

Deflection in the meridian towards the south caused by a deficiency, equal to the mass of the Himalayas, and the mountain region beyond, distributed through a depth of—

	at Kaliána	at Kaliánpur	at Dámargída
100 miles	26″·440	12″·111	6″·855
300 miles	21″·106	11″·678	6″·866
500 miles	17″·066	9″·622	6″·670
1000 miles	11″·199	7″·386	5″·220

Pratt thus showed, that the hypothesis of submontane deficiency was sufficient to produce a considerable amount of compensation for mountain attraction, but by no assumption in the amount of the depth could he reconcile the apparent anomalies in the Indian latitudes. "The "existence of the mountain mass is a fact indisputable", he writes, "not so the compensating "cause, which is simply conjectural as to its existence, and altogether uncertain as to its extent, "if it exist".

Whilst employed on this last calculation it occurred to him that the ocean was another visible cause of disturbance, which might produce a sensible effect. In a paper read before the Royal Society in 1859, (*see* Philosophical Transactions of the Royal Society), he calculates the effect of the ocean on the hypothesis that the deficiency of matter arising from the smaller density of the ocean produces a northerly repulsive force equal to the attraction of a mass of the same form and of a density equal to the excess of the density of rock over that of sea-water.

Pratt assumed that the depth of the ocean was 3 miles, in 28° south latitude, at a point 2,500 miles south of Cape Comorin midway between Madagascar and Australia: at the centre of the Arabian Sea in the latitude of Cape Comorin he took the depth to be 1 mile, and at the centre of the Bay of Bengal in the latitude of Cape Comorin he took the depth to be three-fourths of a mile*. He finds the calculated deflections to be as follows:—

At Cape Comorin	19″·71 North and 2″·19 East.
At Dámargída	10″·44 North and 1″·80 East.
At Kaliánpur	9″·00 North and 0″·48 East.
At Kaliána	6″·18 North and 0″·09 East.

He points out that although many depths in the Atlantic of 4 miles have been measured, no deep sea soundings had been taken in the Indian or Southern oceans, and that his results must therefore be regarded as demonstrating that the effect of the ocean is of importance, rather than as determining its amount.

In a paper on the Indian Arc of Meridian (Philosophical Transactions of the Royal Society, 1860) he abandons the theory, which he had advanced in all his previous papers, that the Indian Arc was curved differently to the mean meridian of the earth, and he finally puts forward the explanation, that *an excess of density in subterranean matter not far south of Kaliánpur will deflect*

* Modern Admiralty charts give Soundings which show that Pratt's depths were not too great.

the plumb-line at Kaliánpur to thes outh, and will account for the anomalies in the results of the Indian Observed Latitudes. In a pamphlet, published in 1869, at Calcutta, he speaks of peculiarities in local attraction, "such as that near Moscow, and this near Kaliánpur", with a certainty greater than is visible in his paper of 1860. In his Figure of the Earth, dated 1871, he calculates that there is a *southerly* deflection of the plumb-line of 3″·55 at Kaliánpur, and he argues that an excess of density *must* exist *near* Kaliánpur, "for the deflection at Kaliánpur", he writes, "is *considerably south*" *.

The zone of Positive Values.

Pratt's calculations were based on but three observed latitudes: we now possess the results of 161 observed latitudes, and cannot solve Pratt's problem. In the following table, are arranged the results of all the Indian astronomical latitudes brought up to date, 1901; reflection will show that the value of (O — C) is equal to the difference between the local attraction at any station and the local attraction at Kaliánpur: assuming that the latter is nothing, a positive value of (O — C) denotes southerly attraction, and a negative value northerly. The persistence of the negative sign is noticeable, not only under the Himalayas but over Southern India. The results of the Kaliánpur group having rendered problematical the *southerly* deflection of 2″ at Kaliánpur—a deflection which had been assumed in order to account for the excess of negative values of (O — C) in India—I have endeavoured to consider on what other hypothesis the excess of negatives can be explained. From the accompanying table it will be seen, that there are 117 instances of negative excess and 44 instances of positive excess: if we analyse the 44 instances of positive excess, we find that nine occur on or near the coast line from Madras to Mangalore: that five, mostly small in amount, occur in isolated positions amongst surrounding negatives in central India; that three, two of which are less than 1″, occur near Amritsar in Northern India, and that 27 cases occur within a belt or zone, crossing India from Karachi to Calcutta. (*Vide* Chart No. 6).

If the plumb-line throughout India were attracted to the north by some great excess of matter, situated beyond our northern frontier, the values of (O — C) north of the parallel of Kaliánpur would be negative, and the values of (O — C) south of Kaliánpur positive: this is clear, because the value of C depends on the deflection at Kaliánpur, and this deflection, if the attracting force is north, is less than the deflection at northern stations and greater than the deflections at southern stations.

If on the other hand the plumb-line throughout India were repelled to the north by some great deficiency of matter situated south of Cape Comorin, the values of (O — C) south of the parallel of Kaliánpur would be negative and the values of (O — C) north of Kaliánpur positive.

If we suppose the northern attraction and southern repulsion to be acting simultaneously, we find that the northern force produces negative values north of Kaliánpur and positive south: the southern force produces negative values south of Kaliánpur and positive north: at a certain parallel of latitude, dependent on the relative rates of decrease of the two effects, the resultant deflections will equal the deflections on the parallel of Kaliánpur, and the values of (O — C) will equal nothing.

General Walker's figures in *India's Contributions to Geodesy* show that the substitution of Clarke's Axes for Everest's increases the negative tendency of (O — C) in latitudes south of Kaliánpur, and the positive tendency north of Kaliánpur: this substitution will accentuate and not remove the zone of positives.

---

* Colonel Clarke, in his calculations of the Figure of the Earth, also makes the deflection at Kaliánpur to be *south* by 3″·578 in the *Account of the Principal Triangulation of the Ordnance Survey*, by 3″·678 in *Volume XXIX Memoirs R.A.S.*, by 1″·392 in his *Geodesy*. The results of the group of observations have thus falsified all theoretical predictions.

Table showing the Differences between

Meridian of 67°

Station	Lat.	O−C
Karachi ...	24 50	− 0·42

Meridian of 69° and 70°

Station	Lat.	O−C
Khori ..	25 1	− 1·8
Chánga ...	24 59	− 0·1
Rojhra ...	24 57	− 0·4
Alamkhán	24 49	− 1·5

Meridian of 71° and 72°

Station	Lat.	O−C
Dera Dín Panáh	30 34	− 2·56
Mooltán ...	30 11	− 2·78
Ládimsir ...	29 22	− 2·05
[illegible]	[illegible]	[illegible]
Lánki ...	24 58	− 4·5
Virária ...	24 57	− 3·8
Didáwa ...	24 51	− 2·3
[illegible]	[illegible]	[illegible]

Meridian of 73°

Station	Lat.	O−C
[illegible]	[illegible]	[illegible]
[illegible]	[illegible]	[illegible]
[illegible]	[illegible]	[illegible]
Thob ...	26 3	− 3·22
Samdari ...	25 49	− 0·25
Guru Sikkar	24 38	− 3·47
Deesa ...	24 15	− 8·48
Chaniána ...	24 7	−11·50
Sonáda ...	23 7	− 4·56
Colába ...	18 54	−10·64

Meridian of 74° and 75°

Station	Lat.	O−C
Murree ...	33 55	−20·31
[illegible]	[illegible]	[illegible]
[illegible]	[illegible]	[illegible]
[illegible]	[illegible]	[illegible]
Khimúána ...	30 22	− 3·39
Sawaipur ...	29 39	− 1·14
Rám Thal ...	28 29	− 0·77
Garinda ...	27 55	− 0·81
[illegible]	[illegible]	[illegible]
[illegible]	[illegible]	[illegible]
Khámor ...	25 45	− 4·32
Aramlia ...	24 25	− 4·92
Deo Dongri	23 26	− 4·93
Harnása ...	22 47	− 3·51
[illegible]	[illegible]	[illegible]
Valvádi ...	20 44	− 6·77
Dhaigaon ...	19 30	− 4·53
Khánpisura	18 45	− 8·30
Mándvi ...	18 38	− 3·33
[illegible]	[illegible]	[illegible]
Majala ...	16 47	− 1·68
Mávinhúnda	16 25	− 0·03
Navalúr ...	15 25	− 3·00
Kundgol ...	15 15	− 1·13
Hónnavalli ..	14 16	− 2·01
Koramúr ...	14 8	− 5·19
[illegible]	[illegible]	[illegible]

Meridian of 76°

Station	Lat.	O−C
Isanpur ...	30 38	− 4·29
Rákhi ...	29 17	− 0·83
Tásing ...	27 53	− 0·29
Bánskho ...	26 50	− 5·83
Kánkra ...	25 38	− 1·09
Gurária ...	24 26	− 0·79
Kanheri ...	18 29	− 9·12
Nitali ...	18 17	− 4·67
Kem ...	18 11	− 3·53

Meridian of

Station
Banog ...
Mussooree ...
Rajpur ...
Amsot ...
Dehra Dún ...
E. End Dehra Base
Nojli ...
Kaliána ...
Datairi ...
Bostán ...
Chandaos ...
Noh ...
Agra ...
Usira ...
Kesri
Pahárgarh ...
Daicdhari ...
Surantál
Sironj
N. E. End Base
Bhaorása
Kaliánpur ...
Losalli ...
Tinsia
Kámkhera ...
Ahmadpur ...
Ládi ...
Takalkhera ...
Badgaon ...
Voi ...
Achola ...
Dámargída ...
Kodangal ...
Tuagat ...
Namthabad ...
Hónnur
Pavagada
Bangalore ...
Pachapálaiyam
Kutipárai ...
Punnæ

The Latitude of Kaliánpur has been taken 24° 7′ 11″·57. **Positive Values are printed in red.**

Observed and Computed Values of Latitude.

77° 30′

Lat. ° ′	O−C
30 28	−33·04
30 27	−36·84
30 24	−47·65
30 22	−29·15
30 19	−37·82
30 17	−30·37
29 53	−13·95
29 31	− 7·03
28 44	− 6·13
28 31	− 5·67
28 5	− 1·16
27 51	− 0·26
27 10	− 5·72
26 57	− 6·03
25 40	+ 5·45
24 59	− 0·70
23 58	+ 1·01
24 14	+ 0·82
24 9	+ 1·09
24 8	+ 1·17
24 7	
24 6	− 1·02
24 6	+ 0·98
24 0	− 2·15
23 16	− 2·40
23 5	− 5·34
22 6	− 6·90
20 44	− 7·83
19 7	− 5·51
18 15	− 3·52
18 3	− 2·74
17 8	− 3·92
16 10	− 5·24
15 6	− 0·96
[illegible] 55	+ 2·93
[illegible] 0	+ 3·10
13 1	− 4·16
[illegible] 0	+ 0·96
[illegible] 29	+ 2·25
[illegible] 9	+ 2·01

Meridian of 78° 30′

Station	Lat. ° ′	O−C
Sarkára ...	29 16	−12·11
Sirsa ...	28 54	− 9·65
Bánsgopál	28 33	− 5·07
Sankráo ...	28 2	− 0·35
Salímpur ...	27 47	− 0·51
Pirmulo ...	17 53	− 4·78
Vánákonda	17 30	− 6·89
Bolarum ...	17 30	− 6·22

Meridian of 80°

Station	Lat. ° ′	O−C
Rámnápur ...	28 22	−11·31
Jarúra ...	28 0	− 0·10
Nimkár ...	27 21	− 0·34
[illegible]	[illegible] 54	+ 4·44
[illegible]	[illegible] 16	+ 5·17
[illegible]	[illegible]	+ [illegible]
[illegible]	[illegible] 27	+ [illegible]
Potenda ...	24 37	+ 1·21
Amua ...	24 0	+ 0·55
Rangír ...	24 0	− 1·40
Lora ...	23 32	+ 4·45
[illegible]	23 11	+ 4·74
Sarandi Pat ...	22 13	− 5·21
Lingmára ...	21 43	− 7·97
Sítápár ...	21 25	− 6·98
Bhimsain ...	20 57	− 7·68
Rájuli ...	20 13	− 4·47
Díwai ...	19 49	− 5·98
Ankora ...	19 25	− 8·38
Burgpaili ...	18 54	− 4·00
[illegible] ...	18 55	+ 0·51
Bolíkonda ...	17 43	− 7·01
Niálamari ...	17 1	− 7·98
Dhúlipaila ...	16 26	− 3·58
Dánapa ...	15 50	− 0·75
Ongole ...	15 30	− 4·26
Darutippa ...	15 1	− 3·26
Kistama ...	14 27	− 2·57
Gudah ...	14 1	+ 0·89
Madras ...	13 4	+ 4·66
Tiruvendipuram	11 45	+ 5·45

Meridian of 81° and 82°

Station	Lat. ° ′	O−C
Karára ...	24 5	− 0·12
Gurwáni ...	24 1	+ 2·91
Dalea ...	22 19	− 3·62
Patháídi ...	21 49	− 3·17
Ramai ...	20 57	− 1·42
[illegible] ...	19 52	+ 0·05
Karía ...	19 12	− 3·55
Singáwáram	17 45	− 2·00
Sánjib ...	17 31	− 6·07
Parampúdi ...	17 13	− 5·97

Meridian of 84°

Station	Lat. ° ′	O−C
[illegible] ...	24 2	+10·44
Khundábolo ...	19 51	− 5·88
Mal ...	18 47	−10·53
Ráwal ...	18 32	− 4·89
Vizagapatam Base	18 1	− 6·62
Waltair ...	17 43	− 9·24

Meridian of 86°

Station	Lat. ° ′	O−C
Chendwár ...	23 57	+ 2·70
Dariápur ...	21 47	+ 0·54
Cuttack ...	20 29	− 8·96

Meridian of 88°

Station	Lat. ° ′	O−C
[illegible]	22 54	+ 5·31
Calcutta	22 33	+ 0·60
Patna	21 47	− 3·79
Chandípur	21 27	− 3·37

The horizontal black lines show the limits of the zone where positive values predominate.

The apparent zone of positives would not be eliminated, even if a southerly deflection of 2″ at Kaliánpur were proved. If the values of (O – C) in all India be corrected by + 2″, the *difference* in the mean direction of gravity in *the zone* from that in all *India* will *remain the same*: there would still be a greater tendency to southerly deflection within the zone than beyond it.

Mean value of (O – C) in the zone of positives ... ... = + 1″·04
Mean value of (O – C) south of the zone of positives ... = – 3″·67
Mean value of (O – C) north of the zone of positives ... = – 9″·48
Mean value of (O – C) north of the zone, if six large Himalayan values are excluded ... ... = – 3″·66
Mean value of (O – C) in latitude for all India ... ... = – 3″·83
Mean value of (O – C) for all India excluding six large values = – 2″·47

Whatever correction be applied to the initial latitude to eliminate the value – 3″·83 for all India, the direction of gravity *within* the zone will remain inclined to the mean meridional direction for all India by 4″·87.

Results of the Indian Observed Longitudes.

There are 24 Longitude stations in India and Burma.

The following table gives the values of (O – C) in Longitude for the arcs connecting Kaliánpur with each station.

	(O – C)	Deflection in Prime Vertical = (O – C) cos φ
(a) Stations near the meridian of Kaliánpur.	″	″
Amritsar	– 3 ·00	2 ·55 West
Agra	+ 5 ·55	4 ·94 West
Bellary	+ 0 ·75	0 ·72 East
Bolarum	– 3 ·45	3 ·29 East
Bangalore	+ 2 ·85	2 ·78 East
Nágarkoil	+ 1 ·80	1 ·78 East
(b) Stations in Western India.		
Bombay	– 6 ·75	6 ·39 West
Mangalore	– 1 ·95	1 ·90 West
Mooltan	+ 5 ·10	4 ·41 East
Karachi	+ 0 ·45	0 ·41 East
Deesa	+ 3 ·60	3 ·28 East
(c) Stations in Eastern India.		
Fyzabad	– 0 ·45	0 ·40 East
Jubbulpore	–10 ·20	9 ·37 East
Madras	– 7 ·20	7 ·01 East
Waltair]	– 3 ·30	3 ·14 East
Calcutta	– 10·95	10·12 East
(d) Stations in Burma.		
Chittagong	–11 ·70	10·82 East
Akyab	–11 ·70	10·99 East
Prome	–16 ·35	15·48 East
Moulmein	–18 ·30	17·55 East
(e) Stations near mountains.		
Peshawar	–14 ·25	11·81 West
Quetta	+ 2 ·40	2 ·07 East
Dehra Dún	–25 ·65	22·14 East
Jalpaiguri	–20 ·40	18·26 East

The preponderance of apparent easterly deflections can be eliminated, if a westerly deflection at Kalíánpur be assumed to exist; as however Captain Lenox Conyngham's observations at the group of surrounding stations do not support this assumption, it cannot be considered justifiable.

The Observed Azimuths of India and Burma.

In Volume II of the Great Trigonometrical Survey of India, General Walker treated 63 observed azimuths in his endeavour to obtain the correct fundamental azimuth of India; many azimuths have been observed since that Volume was written, and a complete list of the observed azimuths of India and Burma is published as Appendix II of this paper. 195 such astronomical azimuths have been observed; the value of (O—C) is positive in 57 cases and negative in 138 cases. But the value of C, the geodetic value, is dependent on General Walker's derived value of the fundamental azimuth, and, before we can treat the observed azimuths as a whole, we must replace the derived value of the azimuth at Kalíánpur by the observed value.

The azimuth at Kalíánpur, as observed at Kalíánpur itself, is 190° 27′ 6″·29, (*vide* page 7), and as deduced from the observations of the surrounding group is 190° 27′ 6″·39, (*vide* page 8); these two values differ but slightly; we will select the latter as the true fundamental Azimuth at Kalíánpur; General Walker's derived Azimuth is 190° 27′ 5″·10, or 1″·29 too small*. If we increase the geodetic azimuths of India by 1″·29, we find that the value of (O — C) becomes positive in 37 cases and negative in 158 cases; General Walker's correction has therefore reduced the number of negative values of (O — C) from 158 to 138, and has increased the number of positive values from 37 to 57. If a line be drawn through Kalíánpur north-east and south-west, *i.e.*, through Bombay, Kalíánpur, Lucknow, it will be found, that 33 of the 37 positive values lie north-west of that line, and of the four values that lie south-east of it three are less than 1″. (*Vide* Chart No. 7).

The triangulation of India was divided for the purpose of simultaneous reduction into six great areas: the first, known as the North-West Quadrilateral, covers the whole country North-West of Kalíánpur; the second, the North-East Quadrilateral, embraces the portion of India North-East of Kalíánpur up to the western Boundary of Burma: the third, the South-East Quadrilateral, is the country South-East of Kalíánpur as far south as latitude 18°: the fourth, the South-West Quadrilateral, covers the country South-West of Kalíánpur as far south as latitude 18°. The fifth, the Southern Trigon, embraces the whole peninsular area south of latitude 18°, and the sixth is the Burma Quadrilateral.

---

* Because the azimuth at Kalíánpur has been re-observed and deduced from the group.

The following table gives the mean value of (O – C) and the numbers of positive and negative values found in the several areas:—

Area	Corrected Mean value of (O – C) in azimuth	General Walker's (O – C)		(O – C) corrected by – 1″·29	
		No. of positive values	No. of negative values	No. of positive values	No. of negative values
	″				
N. W. Quadrilateral ...	– 0·17	36	21	25	32
N. E. Quadrilateral ...	– 6·14	4	24	4	24
S. E. Quadrilateral ...	– 4·53	6	35	3	38
S. W. Quadrilateral ...	+ 0·61	9	2	5	6
Southern Trigon ...	– 6·62	2	34	0	36
Burma Quadrilateral ...	–13·14	0	22	0	22
	Total ...	57	138	37	158

No positive values occur in Burma or the Southern Trigon: the few that fall in the North-East Quadrilateral, the South-East Quadrilateral and the South-West Quadrilateral are either north-west of the dotted line in Chart No. 7, or situated very near to it on the southern side.

It will be instructive to compare the deflections of the plumb-line in the Prime Vertical, as deduced from Longitude and Azimuth observations, at those stations where both sets of observations have been taken:

Area	Station	Deflection in Prime Vertical		Discrepancy between the Values	Mean Discrepancies in the several Areas
		By Azimuth Observations	By Longitude Observations		
		″	″	″	″
N. W. Q.	Karachi	– 6·35 E.	0·41 E.	– 5·94	– 3·41
N. W. Q.	Dehra Dún	– 23·02 E.	22·14 E.	– 0·88	
N. E. Q.	Bisaul (Fyzabad) ...	– 11·12 E.	0·40 E.	– 10·72	– 7·81
N. E. Q.	Ramganj (Jalpaiguri) ...	– 23·15 E.	18·26 E.	– 4·89	
S. E. Q.	Karaundi (Jubbulpore)	– 12·14 E.	9·37 E.	– 2·77	– 8·24
S. E. Q.	Calcutta	– 25·30 E.	10·12 E.	– 15·18	
S. E. Q.	Vizagapatam Base (Waltair)	– 9·90 E.	3·14 E.	– 6·76	
S. T.	Mangalore	– 19·22 E.	1·90 W.	– 21·12	–33·57
S. T.	Bangalore	– 30·11 E.	2·78 E.	– 27·33	
S. T.	St. Thomas's Mount (Madras)	– 26·55 E.	7·01 E.	– 19·54	
S. T.	Kudankulam (Nágarkoil)	– 68·08 E.	1·78 E.	– 66·30	
Burma	Taungzun (Moulmein)...	– 47·53 E.	17·55 E.	– 29·98	–29·98

If the discrepancies in the last column are assumed due to errors in the geodetic azimuths, then the actual errors of the geodetic azimuths can be found by multiplying these discrepancies by the tangent of the latitude thus:—

Area	Mean error in Deflection in Prime Vertical	Corresponding error in Azimuth
	"	"
N. W. Quadrilateral ...	− 3·41	− 1·78
N. E. Quadrilateral ...	− 7·81	− 3·90
S. E. Quadrilateral ...	− 8·24	− 3·21
Southern Trigon ...	− 33·57	− 6·98
Burma	− 29·98	− 8·84

In the following table the mean values of (O — C) in azimuth, obtained by comparing the geodetic and astronomic values of azimuth, are given: and beside them are shown the errors in azimuth, deduced by comparing azimuthal and longitude results.

Area	Mean value of (O−C) in Azimuth obtained by comparing Geodetic and Astronomic Azimuths	Error in Azimuth deduced by comparing the results of Azimuth and Longitude Observations
	"	"
N. W. Quadrilateral ...	− 0·17	−1·78
N. E. Quadrilateral ...	− 6·14	−3·90
S. E. Quadrilateral ...	− 4·53	−3·21
S. W. Quadrilateral ...	...	...
Southern Trigon ...	− 6·62	−6·98
Burma	− 13·14	−8·84

It will be seen that there are reasons for believing, that the persistence of the negative sign in the azimuthal values of (O — C) is due to errors of triangulation. In the North-West Quadrilateral the deduced error in Azimuth is 2″ in excess of the mean value of (O — C): in the Southern Trigon the mean value of (O — C) agrees with the deduced error in Azimuth: in the North-East and South-East Quadrilaterals, the deduced error is 2″ less than the value of (O—C): in Burma the deduced error of the triangulation is — 9″ and the value of (O — C) is — 13″.

The triangulation of the North-West Quadrilateral was the first reduced, and that of the South-East Quadrilateral followed: the North-East Quadrilateral had thus its western and southern sides fixed when its reduction began; The Southern Trigon had the eastern half of its northern side fixed before its reduction, and the South-West Quadrilateral was fitted in between three fixed sides. Any error in azimuth in the triangulation of the North-West Quadrilateral and South-East Quadrilateral will therefore affect the North-East Quadrilateral: the errors of the South-East Quadrilateral will be carried into the Southern Trigon, and those of the latter into the southern half of the South-West Quadrilateral: the errors of the triangulation of the North-East Quadrilateral will be carried on into Burma. The mean azimuthal closing errors of circuits of triangulation of the six areas are as follows:—

Area	Mean Closing Error in Azimuth of the Triangulation	Average Error in Azimuth generated in 10 triangles of Triangulation
	"	"
N. W. Quadrilateral ...	0·7	0·28
N. E. Quadrilateral ...	0·8	0·83
S. E. Quadrilateral ...	2·9	0·25
S. W. Quadrilateral ...	3·2	1·33
Southern Trigon ...	1·8	0·48
Burma	2·25	0·58

Positive values of (O – C) occur both in the North-East Quadrilateral and the South-West Quadrilateral along the lines, where these two areas abut against the North-West Quadrilateral, and doubtless the correctness of the latter's orientation has made itself felt for some distance within the interiors of the abutting areas: the occurrence of these positive values has tended to reduce the mean value of (O – C) in both the North-East Quadrilateral and the South-West Quadrilateral. Burma is affected by the full force of the errors of the North-East Quadrilateral; the closing error of the eastern circuit of the North-East Quadrilateral, the circuit to which Burma is attached, is – 13″·14.

Though the difference in azimuth between two rays can be more accurately determined by triangulation than by astronomical observations when the rays are not distant from one another, yet the errors of triangulation tend to accumulate and at great distances from the origin such as Cape Comorin or Moulmein, the accumulated error of the triangulation may exceed the error that local attraction is liable to produce in an observed azimuth.

Other possible causes of the preponderance of negative values of (O – C) in Azimuth.

It has been shown that there are *reasons for believing*, that the persistence of the negative sign *may* be due to errors that have accumulated in the triangulation. But until the uncertainty, which at present surrounds the subject of Himalayan attraction, has been removed, no conclusion can be final. In the Table on page 17 it had to be assumed that the errors of the axes of the Everest spheroid had the same effect in both Longitude and Azimuth: it was assumed, in fact, that the discrepancies between the Longitude and Azimuthal results were not due to errors of spheroid: this assumption is not justifiable; it may be correct, but we have no present means of telling.

At stations, however, on the meridian of Kaliánpur, the errors of the adopted spheroid have no effect in either longitude or azimuth; and yet the deflection in the prime vertical at Kudankulam (Nágarkoil) as deduced from longitude observations is 1″·78 East, and from azimuth observations is 68″·08 East. If the island of Ceylon attracts the plumb-line at Kudankulam (Nágarkoil) or if the Arabian Sea repels it, the same effect should be exhibited by the longitude and azimuth observations. It is difficult to avoid the conclusion that the discrepancy here is due to errors of the triangulation.

The largest values of (O – C) in azimuth occur in Burma, where the negative sign is persistent. The Himalaya mountains tend to render values of (O–C) in Burma positive: their effect, if they have any, is therefore masked by more powerful influences. The interposition of the Bay of Bengal between India and Burma gives a positive tendency to values of (O–C) in India and a negative tendency to Burma, but its presence does not account for the discrepancy between the results of the longitude and azimuth observations at Moulmein, (*vide* page 17). Longitude observations in Upper Burma, and a better knowledge of the heights and masses of the mountains of Burma will help towards the solution of the problem.

Dehra Dún:

June 1900.

POSTSCRIPT.

February, 1901.

A second zone of positive values.

Since the above paper was written, I have come to the conclusion, that the positive values of (O–C) in latitude, that preponderate over the extreme south of India, (*vide* Chart No. 6), may possibly constitute a second zone of positive values, produced like the first by a combination of Himalayan and Oceanic influences. The positive values of South India have always been regarded as proofs, that the plumb-line is attracted *towards* the Ocean*, and on this account great interest attaches to them. If (O–C) in latitude is positive, O must be greater than C, the observed latitude must be too large, the plumb-line must be deflected towards the south: this is the reasoning, that has led us believe in deflections *towards* the Ocean. But in truth we have no justification for assuming C correct. If (O–C) is positive, the only true inference is, that O has been influenced less by northerly attraction (or more by southerly attraction) than C.

At Punnæ in the south of India the value of (O–C) in latitude is positive: the inference has been drawn that the plumb-line at Punnæ is deflected towards the Ocean. If we are correct in arguing from the preponderance of negative values throughout India, that *the Deflection at Kaliánpur is south*, then the positive value of (O–C) at Punnæ denotes southerly attraction. But if we pretend to no knowledge of the absolute deflection at Kaliánpur, then the positive value at Punnæ merely indicates a more southerly or less northerly deflection than at Kaliánpur.

The positive value of (O–C) at Amritsar near the Himalayas, (*vide* meridian of 74° in Table following page 14), has been held to prove, that Himalayan attraction has no far-reaching effect: but this proof again is dependent on the correctness of the method of deducing the deflection at Kaliánpur. If the deflection at Kaliánpur as deduced from the results of all India is accepted, then the positive value at Amritsar proves the weakness of Himalayan attraction.

---

* *Vide* page 806, Philosophical Transactions Royal Society, Volume 186, 1895, India's Contributions to Geodesy: "The whole of the arcs (*i.e.*, of longitude), except those from Waltair, show deflection towards the ocean and not towards "the interior of the continent. The astronomical latitudes in the Southern Peninsula tell the same tale of deflection "towards the ocean".

But the only true inference from the results of the group is, that the meridional deflection at Amritsar is *less northerly* than at Kaliánpur, a fact that is not surprising, seeing that the Himalayan mass is east of Amritsar and north-east of Kaliánpur.

Examples might be multiplied, but it is only necessary to mention one more, *viz.*, the case of Kesri, on the meridian of 77° 30′ in latitude 25° 46′, (*vide* the Table following page 14). Kesri is 112 miles due north of Kaliánpur, and the appearance of a large positive value between Kaliánpur and the Himalayas has been considered to indicate the absence of Himalayan influence: but this positive value has no real significance: it merely denotes, that some local cause gives a deflection to the plumb-line at Kesri, more southerly by 5″, than the deflection at Kaliánpur. The Kesri result will continue to denote this single fact, whether Himalayan attraction is found to be far-reaching or not*.

Let us assume that the alternate positive and negative zones, shown on Chart No. 6, are due to the combined influences of the Himalayas and the Ocean: then the positive zone in latitude 25° will signify, that northwards from Kaliánpur the influence of the Ocean is decreasing more rapidly than that of the Himalayas is increasing: as however the Himalayas are more nearly approached, their influence begins to increase more rapidly: in the centre of the positive zone there will be a line of maximum positive values of (O — C), and of minimum absolute deflections; from this line the increase in the Himalayan influence is greater than the diminution of the Oceanic influence, and in latitude 26° a line is met with, along which the deflections are again as great as that of Kaliánpur: thenceforward northwards the deflections increase rapidly.

Southwards from Kaliánpur the preponderance of negative values denotes that the influence of the Ocean is increasing to the south more rapidly than that of the Himalayas is decreasing: between the parallels of 18° and 20° a belt of maximum negative values is found to exist, which indicate the places of the greatest northerly deflections south of Kaliánpur: after this belt is passed, Himalayan attraction begins to decrease more rapidly than Oceanic influence is increasing: deflections consequently become less northerly south of latitude 18°, and in latitude 14° are again equal to the deflection at Kaliánpur: south of latitude 14°, deflections continue to decrease owing to the waning effect of the Himalayas, and a second positive zone is created. The slow rate, at which the Oceanic influence increases southwards from latitude 18°, is in strange contrast to the rapid increase of deflections, that accompanies an approach to the Himalayas, and may possibly be due to the fact that the effects of the Arabian Sea and the Bay of Bengal south of latitude 15° begin to oppose the influence of the Indian Ocean.

If we divide India into the four zones of Chart No. 6, the negative and positive values of (O — C) in latitude will be found distributed as follows:—

		Positive Values	Negative Values
(1)	Northern negative zone	3	33
(2)	Positive zone north of Kaliánpur ...	26	13
(3)	Negative zone south of Kaliánpur ...	5	70
(4)	Southern positive zone	9	1

The mean values of (O — C) in latitude are as follows:—

		″
(1)	Within the northern negative zone	— 9·48
(2)	Within the northern positive zone	+ 1·04
(3)	Within the southern negative zone	— 4·47
(4)	Within the southern positive zone	+ 2·08
(5)	In all India	— 3·85

* The surface of India consists of alluvium north of Kesri, and of rock to the south: the change from alluvium to gneiss, trap and Vindhyan rock occurs near Kesri. East of Kaliánpur the northern positive zone follows the line, where the alluvium and rock join, as far almost as the Bay of Bengal.

No mean correction, applied to the latitude of Kaliánpur, can alter the *differences* between the mean values of (O – C) of the several zones: a diminution of 2″ in the latitude of Kaliánpur will decrease the areas and mean values of the negative zones, and increase those of the positive zones.

The existence of the zones cannot be attributed to errors of the Everest spheroid: if we select a spheroid that eliminates a positive preponderance north of Kaliánpur and a negative preponderance south, we enhance the apparent effects of Himalayan attraction in the northern negative zone, and increase the positive values in South India: we also cause the northern negative zone to expand southwards, and the southern positive zone to spread northwards. If we select a spheroid that eliminates the positive values in South India, we cover the whole Indian peninsula south of Kaliánpur with negative values, and we increase the northern positive zone.

S. G. B.

PART II.

The deflections at Kalianpur calculated from the configuration of the ground in the vicinity.

The two rival systems of deflections.

In Charts Nos. 4 and 5 are given four diagrams; two of these diagrams, Nos. 1 and 3, show the deflections of the plumb-line as deduced from the results of the Kalíánpur group: the remaining two diagrams Nos. 2 and 4 show the deflections of the plumb-line as deduced from the results of all India. Diagrams Nos. 1and 3 have been constructed from the astronomical data derived from a group of stations, all situated within a radius of 35 miles: Diagrams Nos. 2 and 4 have been formed from data derived also from a group of stations, the area embracing all India. There are no theoretical reasons for limiting the area of a group, and there are no scientific objections to a large group. But if an area of 3,000 square miles were bounded on the north by high mountains and on the south by deep seas, it would not be considered a favorable locality for an *absolute* determination of deflection; and an area of 4,000,000 square miles, if similarly situated, may possibly be unfavorable also.

When considering whether the first or the second of the two rival systems of deflections is the more probable or the more acceptable, we must never lose sight of the essential difference between them: Diagrams Nos. 1 and 3 merely pretend to show the *relative* deflections of the plumb-lines *inter se* at the stations of the Kalíánpur group: the plumb-lines may all have a large constant additional deflection superimposed by external sources of disturbance; but of external sources Diagrams Nos. 1 and 3 have no cognisance: Diagrams Nos. 1 and 3 *give the absolute deflections, that would obtain at Kalíánpur, if all outside influences were removed:* they give, in fact, *deflections due to purely local attraction.*

Diagrams Nos. 2 and 4 go further: they pretend to give not relative deflections but absolute. Diagram No. 2 states definitely that the plumb-line at Kalíánpur is deflected to the south through 2″. Diagram No. 1 claims no knowledge of Himalayan attraction: Diagram No. 2 certifies, that neither the Himalayas nor the Ocean have any influence at Kalíánpur. When the negative values of (O—C) in latitude and azimuth predominate in India over the positive, it is easy to bring about an equality, if we diminish the latitude and azimuth of our station of origin: but this expedient entails the assumption of a knowledge of Himalayan attraction greater than we possess.

The two systems of deflections may be exhibited thus:—

Station	In the Meridian		In the Prime Vertical	
	The "Group" System	The "Mean of India" System	The "Group" System	The "Mean of India" System
	"	"	"	"
Daiádhari ...	+ 1·01 S.	+ 3·61 S.	+ 2·13 W.	+ 5·35 W.
Súrantál ...	+ 0·82 S.	+ 3·42 S.	+ 3·64 W.	+ 6·86 W.
Sironj, N.E. End Base	+ 1·69 S.	+ 4·29 S.	+ 2·54 W.	+ 5·76 W.
Bhaorása ...	+ 1·17 S.	+ 3·77 S.	+ 0·22 W.	+ 3·44 W.
Kaliánpur ...	− 0·60 N.	+ 2·00 S.	− 0·22 E.	+ 3·00 W.
Losalli ...	− 1·02 N.	+ 1·58 S.	− 6·38 E.	− 3·16 E.
Tinsia ...	+ 0·98 S.	+ 3·58 S.	...	...
Salot ...	...	...	− 4·49 E.	− 1·27 E.
Kámkhera ...	− 2·15 N.	+ 0·45 S.	+ 0·04 W.	+ 3·26 W.
Ahmadpur ...	− 2·49 N.	+ 0·11 S.	+ 2·27 W.	+ 5·49 W.

Whatever value be adopted for Kaliánpur itself, its difference from the deflection of each other station will remain the same: therefore, if we impose on Kaliánpur a deflection of 2″ to the south in the meridian*, we *must* increase the deflection of every station, as given by the group system, by 2″·60. *A deflection of 2″ to the south at Kaliánpur must be accompanied by a southerly deflection at every station of the group.* There should exist therefore some powerful hidden cause, affecting the whole area of the group and vitiating all observations; it should not be far from Kaliánpur, as it does not affect India as a whole, its existence having been assumed for the purpose of equalising positive and negative values throughout the peninsula. I put forward the plea, that we should locate this cause *in situ*, before we use it in support of theories. If attractions were due only to visible hills, it would not be possible for us to assume suitable deflections, unless they were justified by the actual configuration of the ground: an imaginary subterranean cause is not a safe explanation of theoretical anomalies, unless it be accompanied by direct proof from the ground.

It had been prophesied that Kaliánpur would be found to resemble Moscow. At Moscow on a line 60 miles long, running nearly east and west over a flat plain, northerly deflections of 8″

* For the purpose of equalising the numbers of positive and negative values of (O − C) in latitude.

are found: along a parallel line nine miles to the south the plumb-lines hang vertical, and along a third line, nine miles further to the south, the deflections are southerly and amount to 8″. Fourteen miles north of the line of maximum northerly attractions, and fourteen miles south of the line of maximum southerly attractions the plumb-lines recover the vertical position. A case such as Moscow, though often quoted as typical, is practically unique: there is nothing in India to be compared with it.

Before we can investigate the deflections at Kaliánpur, that are due to *subterranean* causes, we must clear the *observed* deflections of the effects of the *visible* configuration of the surface. A description of the country round Kaliánpur is given in Appendix I of this paper. The surface is flat and there are no mountains: but a plateau exists 200 to 250 feet higher than the general level of the country, and some of the stations of the group are situated on the plateau, others to the south and east of it. The deflections due to this plateau and to surrounding dislevelments of the surface must be calculated.

The Method of calculation.

The method of calculation has been taken from Colonel Clarke's Geodesy. We possess maps of the district on the scale of one inch to the mile. Round each station, as a common centre, ten circles have been drawn, and through each station a series of twelve radial lines: the country round each station has been thus divided into a series of four-sided compartments: let a_1 and a' be the azimuths of two consecutive lines, and r_1 and r' the radii of two consecutive circles, then Colonel Clarke shows that the deflection in the direction north caused by the mass of the compartment contained between limits a_1 and a', and r_1 and r' is

$$12''{\cdot}44 \frac{\delta}{\Delta} h \, (\sin a' - \sin a_1) \log_e \frac{r'}{r_1},$$

where δ is the density of the mass, Δ the mean density of the earth, and h the average height of the upper surface of the mass above the station.

The radial lines have been drawn at equal intervals of 30° in azimuth: the method, followed in the Ordnance Survey, of so drawing these lines, that the sines of their azimuths were in arithmetic progression, could not be adhered to, as the deflections both in the prime vertical and in the meridian were required at Kaliánpur. The approximate deflection in the prime vertical was derived from the formula

$$12''{\cdot}44 \frac{\delta}{\Delta} h \, (\cos a' - \cos a_1) \log_e \frac{r'}{r_1}.$$

The radius r' was taken equal to $2r_1$, and thence $\log_e \frac{r'}{r_1}$ is equal to 0·693.

The deflection due to each sector and not to each compartment was calculated: $\frac{\delta}{\Delta}$ was assumed $= 0{\cdot}5$: the formula for the deflection in the meridian for each sector thus became

$$12''{\cdot}44 \times 0{\cdot}5 \times 0{\cdot}693 \times \frac{[h] - 9H}{5280} \times (\sin a' - \sin a_1)$$

$$= 0''{\cdot}000817 \, \{[h] - 9H\} \, (\sin a' - \sin a_1),$$

where H = the height of the station, and $[h]$ = the sum of the average heights of the nine compartments in a sector.

The calculation was extended to a distance of 64 miles from each station: this limit was adopted, because irregular masses of ground situated at a greater distance than 64 miles will not affect Kaliánpur *differently* to the mean of the group: our object was to find not the absolute deflection at Kaliánpur resulting from all external causes, but the relative deflections existing at stations of the group, and we could therefore neglect all distant masses, whose average effect on the group did not differ from their effect at Kaliánpur itself.

The calculation was commenced at 220 yards from each station: the maps are not contoured and relative heights cannot be ascertained with sufficient accuracy to justify a nearer approach than 220 yards. The heights of compartments were read off the map, and were averaged for this calculation by Captain Lenox Conyngham, whose intimate knowledge of Kaliánpur and its vicinity enabled him to appreciate the topographical features of the map. The ground immediately surrounding the stations was as a rule flat and without feature: at the few stations, where there was a drop of 100 feet within the radius of 220 yards, the drop was uniform on each side. The only station, about which Captain Lenox Conyngham felt uncertainty, was Ahmadpur: this station is situated on a truncated pyramid of rock, rising 250 feet above the plain: a drop of 150 feet occurs on each side within 220 yards, and at the summit the station is nearer to the southern precipice than to the northern slope. The error caused in the calculated value of the deflection by this want of symmetry may be found approximately as follows:—Suppose the top of the hill to be a circle of 30 yards in diameter, and suppose the station to be 10 yards from the southern precipice and 20 yards from the northern: then on the north side there will exist part of a hollow cylinder concentric with the station: the inner radius of this cylindrical shell is 10 yards, and the thickness of the shell is 10 yards: the walls of the shell do not extend south of the prime vertical. The effect of such a shell on the north side, if uncompensated by matter on the south side, will be by Colonel Clarke's formula, assuming the height of the shell to be 150 feet,

$$12''\cdot 44 \times \tfrac{1}{2} \times \frac{150}{5280} \times 0\cdot 693 \times 2 = 0''\cdot 24.$$

DAIÁDHARI.

Height above Mean Sea Level = 1867 feet.

Heights of Compartments in feet.

Radii of Annuli		SECTORS											
		N.	N.E.	E.		S.E.	S.		S.W.	W.		N.W.	N.
		$\alpha' = 30°$	$= 60°$	$= 90°$	$= 120°$	$= 150°$	$= 180°$	$= 210°$	$= 240°$	$= 270°$	$= 300°$	$= 330°$	$= 0°$
r'	r_1	$\alpha_1 = 0°$	$= 30°$	$= 60°$	$= 90°$	$= 120°$	$= 150°$	$= 180°$	$= 210°$	$= 240°$	$= 270°$	$= 300°$	$= 330°$
miles 0·25	miles 0·125	1700	1700	1750	1720	1720	1720	1700	1700	1700	1700	1700	1700
0·5	0·25	1700	1700	1730	1700	1700	1700	1700	1700	1700	1700	1700	1700
1	0·5	1700	1700	1700	1700	1700	1700	1700	1700	1700	1700	1700	1700
2	1	1700	1680	1680	1680	1680	1700	1700	1700	1700	1700	1700	1700
4	2	1680	1670	1640	1660	1660	1680	1700	1700	1700	1700	1700	1650
8	4	1640	1640	1600	1640	1620	1650	1700	1700	1700	1700	1650	1600
16	8	1600	1620	1540	1450	1500	1650	1700	1700	1700	1720	1500	1500
32	16	1600	1550	1400	1400	1400	1680	1700	1700	1500	1550	1600	1400
64	32	1550	1500	1400	1400	1500	1400	1700	1400	1300	1200	1300	1500
Sum − 16803 = S		−1933	−2043	−2363	−2453	−2323	−1923	−1503	−1803	−2103	−2133	−2253	−2353
S × ·000817 = R		−1·579	−1·669	−1·931	−2·004	−1·898	−1·571	−1·228	−1·473	−1·718	−1·743	−1·841	−1·922
$\sin\alpha' - \sin\alpha_1 = A$		+0·500	+0·366	+0·134	−0·134	−0·366	−0·500	−0·500	−0·366	−0·134	+0·134	+0·366	+0·500
R × A = Deflections in Meridian		−0·79	−0·61	−0·26	+0·27	+0·69	+0·79	+0·61	+0·54	+0·23	−0·23	−0·67	−0·96
$\cos\alpha' - \cos\alpha_1 = B$		−0·134	−0·366	−0·500	−0·500	−0·366	−0·134	+0·134	+0·366	+0·500	+0·500	+0·366	+0·134
R × B = Deflections in Prime Vertical		+0·21	+0·61	+0·97	+1·00	+0·69	+0·21	−0·16	−0·54	−0·86	−0·87	−0·67	−0·26

Calculated Total Deflection in the Meridian $= -0''\cdot39$.

Calculated Total Deflection in the Prime Vertical $= +0\cdot33$.

SÚRANTÁL.

Height above Mean Sea Level = 1802 feet.

Heights of Compartments in feet.

Radii of Annuli		SECTORS											
		N	N.E.	E.		S.E.	S.		S.W.	W.		N.W.	N.
		$a' = 30°$	$= 60°$	$= 90°$	$= 120°$	$= 150°$	$= 180°$	$= 210°$	$= 240°$	$= 270°$	$= 300°$	$= 330°$	$= 0°$
r'	r_1	$a_1 = 0°$	$= 30°$	$= 60°$	$= 90°$	$= 120°$	$= 150°$	$= 180°$	$= 210°$	$= 240°$	$= 270°$	$= 300°$	$= 330°$
miles 0·25	miles 0·125	1750	1780	1760	1760	1760	1760	1780	1760	1760	1760	1760	1760
0·5	0·25	1700	1750	1750	1750	1760	1750	1760	1740	1760	1760	1760	1750
1	0·5	1680	1720	1730	1700	1720	1720	1740	1700	1750	1750	1750	1730
2	1	1680	1650	1700	1650	1700	1680	1720	1650	1750	1750	1750	1720
4	2	1650	1620	1650	1650	1650	1650	1680	1700	1750	1730	1750	1650
8	4	1650	1600	1650	1550	1600	1600	1650	1700	1730	1730	1700	1750
16	8	1600	1550	1500	1450	1450	1550	1690	1720	1700	1750	1650	1700
32	16	1600	1500	1400	1320	1350	1420	1550	1720	1600	1750	1650	1650
64	32	1550	1500	1400	1400	1800	1700	1550	1550	1400	1400	1410	1600
Sum − 16218 = S		−1358	−1548	−1678	−1988	−1428	−1388	−1098	− 978	−1018	− 838	−1038	− 908
S × ·000817 = R		−1·109	−1·265	−1·371	−1·624	−1·167	−1·134	−0·897	−0·799	−0·832	−0·685	−0·848	−0·742
Sin a' − sin a_1 = A		+0·500	+0·366	+0·134	−0·134	−0·366	−0·500	−0·500	−0·366	−0·134	+0·134	+0·366	+0·500
R × A = Deflections in Meridian		−0·55	−0·46	−0·18	+0·22	+0·43	+0·57	+0·45	+0·29	+0·11	−0·09	−0·31	−0·37
Cos a' − cos a_1 = B		−0·134	−0·366	−0·500	−0·500	−0·366	−0·134	+0·134	+0·366	+0·500	+0·500	+0·366	+0·134
R × B = Deflections in Prime Vertical		+0·15	+0·46	+0·69	+0·81	+0·43	+0·15	−0·12	−0·29	−0·42	−0·34	−0·31	−0·10

Calculated Total Deflection in the Meridian = + 0″·11.
Calculated Total Deflection in the Prime Vertical = + 1·11.

NORTH-EAST END OF SIRONJ BASE.

Height above Mean Sea Level = 1481 *feet.*

Heights of Compartments in feet.

Radii of Annuli		SECTORS											
		N.	N.E.	E.		S.E.	S.		S.W.	W.		N.W.	N.
		$\alpha' = 30°$	$= 60°$	$= 90°$	$= 120°$	$= 150°$	$= 180°$	$= 210°$	$= 240°$	$= 270°$	$= 300°$	$= 330°$	$= 0°$
r'	r_1	$\alpha_1 = 0°$	$= 30°$	$= 60°$	$= 90°$	$= 120°$	$= 150°$	$= 180°$	$= 210°$	$= 240°$	$= 270°$	$= 300°$	$= 330°$
miles 0·25	miles 0·125	1470	1470	1470	1470	1470	1470	1470	1470	1470	1470	1470	1470
0·5	0·25	1460	1470	1460	1460	1460	1460	1470	1470	1460	1470	1460	1460
1	0·5	1460	1460	1450	1450	1450	1450	1460	1470	1450	1460	1450	1460
2	1	1460	1450	1440	1440	1440	1440	1450	1470	1450	1450	1450	1450
4	2	1440	1440	1400	1410	1410	1410	1430	1460	1430	1440	1440	1430
8	4	1430	1440	1380	1350	1350	1350	1450	1500	1500	1470	1450	1500
16	8	1400	1400	1300	1320	1300	1380	1430	1600	1650	1630	1600	1520
32	16	1420	1300	1350	1300	1500	1400	1380	1550	1720	1700	1720	1500
64	32	1300	1220	1300	1400	1650	1620	1550	1550	1460	1500	1600	1500
Sum − 13329 = S		− 489	− 679	− 779	− 729	− 299	− 349	− 239	+ 211	+ 261	+ 261	+ 311	− 39
S × ·000817 = R		−0·400	−0·555	−0·636	−0·596	−0·244	−0·285	−0·195	+0·172	+0·213	+0·213	+0·254	−0·032
$\sin \alpha' - \sin \alpha_1 = A$		+0·500	+0·366	+0·134	−0·134	−0·366	−0·500	−0·500	−0·366	−0·134	+0·134	+0·366	+0·500
R × A = Deflections in Meridian		−0·20	−0·20	−0·09	+0·08	+0·09	+0·14	+0·10	−0·06	−0·03	+0·03	+0·09	−0·02
$\cos \alpha' - \cos \alpha_1 = B$		−0·134	−0·366	−0·500	−0·500	−0·366	−0·134	+0·134	+0·366	+0·500	+0·500	+0·366	+0·134
R × B = Deflections in Prime Vertical		+0·05	+0·20	+0·32	+0·30	+0·09	+0·04	−0·03	+0·06	+0·11	+0·11	+0·09	0·00

Calculated Total Deflection in the Meridian $= -0''\cdot07$.

Calculated Total Deflection in the Prime Vertical $= +1''\cdot34$.

BHAORÁSA.

Height above Mean Sea Level = 1387 feet.

Heights of Compartments in feet.

Radii of Annuli		SECTORS											
		N.	N.E.	E.		S.E.	S.		S.W.	W.		N.W.	N.
		$a' = 30°$	$= 60°$	$= 90°$	$= 120°$	$= 150°$	$= 180°$	$= 210°$	$= 240°$	$= 270°$	$= 300°$	$= 330°$	$= 0°$
r'	r_1	$a_1 = 0°$	$= 30°$	$= 60°$	$= 90°$	$= 120°$	$= 150°$	$= 180°$	$= 210°$	$= 240°$	$= 270°$	$= 300°$	$= 330°$
miles 0·25	miles 0·125	1360	1360	1370	1380	1360	1370	1360	1360	1370	1370	1360	1360
0·5	0·25	1340	1340	1360	1370	1340	1380	1350	1370	1360	1360	1350	1350
1	0·5	1340	1340	1350	1370	1340	1360	1380	1370	1360	1350	1340	1340
2	1	1340	1330	1330	1350	1330	1360	1370	1350	1350	1360	1330	1340
4	2	1340	1320	1340	1360	1330	1350	1360	1360	1360	1350	1340	1330
8	4	1320	1320	1330	1350	1350	1320	1330	1330	1380	1380	1380	1350
16	8	1320	1360	1360	1360	1400	1330	1340	1380	1450	1440	1380	1340
32	16	1400	1380	1400	1400	1530	1450	1380	1400	1700	1730	1680	1500
64	32	1300	1200	1300	1400	1650	1650	1600	1550	1600	1500	1600	1500
Sum − 12483 = S		− 423	− 533	− 343	− 143	+ 147	+ 87	− 13	− 13	+ 447	+ 357	+ 277	− 73
S × ·000817 = R		−0·346	−0·435	−0·280	−0·117	+0·120	+0·071	−0·011	−0·011	+0·365	+0·292	+0·226	−0·060
Sin a' − sin a_1 = A		+0·500	+0·366	+0·134	−0·134	−0·366	−0·500	−0·500	−0·366	−0·134	+0·134	+0·366	+0·500
R × A = Deflections in Meridian		−0·17	−0·16	−0·04	+0·02	−0·04	−0·04	+0·01	0·00	−0·05	+0·04	+0·08	−0·03
Cos a' − cos a_1 = B		−0·134	−0·366	−0·500	−0·500	−0·366	−0·134	+0·134	+0·366	+0·500	+0·500	+0·366	+0·134
R × B = Deflections in Prime Vertical		+0·05	+0·16	+0·14	+0·06	−0·04	−0·01	0·00	0·00	+0·18	+0·15	+0·08	−0·01

Calculated Total Deflection in the Meridian = − 0″·38.
Calculated Total Deflection in the Prime Vertical = + 0 ·76.

KALIÁNPUR.

Height above Mean Sea Level = 1765 feet.

Heights of Compartments in feet.

Radii of Annuli		SECTORS											
		N.	N.E.	E.		S.E.	S.		S.W.	W.		N.W.	N.
		$\alpha' = 30°$	$= 60°$	$= 90°$	$= 120°$	$= 150°$	$= 180°$	$= 210°$	$= 240°$	$= 270°$	$= 300°$	$= 330°$	$= 0°$
r'	r_1	$\alpha_1 = 0°$	$= 30°$	$= 60°$	$= 90°$	$= 120°$	$= 150°$	$= 180°$	$= 210°$	$= 240°$	$= 270°$	$= 300°$	$= 330°$
miles 0·25	miles 0·125	1755	1750	1750	1760	1750	1750	1750	1745	1740	1750	1760	1760
0·5	0·25	1735	1720	1720	1750	1740	1740	1740	1740	1730	1720	1750	1750
1	0·5	1700	1690	1720	1750	1720	1720	1720	1700	1720	1710	1750	1750
2	1	1630	1600	1680	1680	1680	1700	1700	1700	1690	1700	1730	1740
4	2	1630	1580	1580	1570	1580	1650	1690	1700	1680	1700	1700	1730
8	4	1680	1460	1450	1450	1500	1750	1650	1650	1680	1680	1700	1700
16	8	1720	1540	1450	1400	1450	1600	1650	1650	1700	1700	1700	1700
32	16	1550	1450	1400	1350	1350	1380	1530	1650	1550	1700	1710	1700
64	32	1580	1500	1400	1400	1800	1650	1500	1500	1350	1400	1410	1600
Sum − 15885 = S		− 905	−1595	−1735	−1775	−1315	− 945	− 955	− 850	−1045	− 825	− 675	− 455
S × ·000817 = R		−0·739	−1·303	−1·417	−1·450	−1·074	−0·772	−0·780	−0·694	−0·854	−0·674	−0·551	−0·372
$\sin\alpha' - \sin\alpha_1 = A$		+0·500	+0·366	+0·134	−0·134	−0·366	−0·500	−0·500	−0·366	−0·134	+0·134	+0·366	+0·500
R × A = Deflections in Meridian		−0·37	−0·48	−0·19	+0·19	+0·39	+0·39	+0·39	+0·25	+0·11	−0·09	−0·20	−0·19
$\cos\alpha' - \cos\alpha_1 = B$		−0·134	−0·366	−0·500	−0·500	−0·366	−0·134	+0·134	+0·366	+0·500	+0·500	+0·366	+0·134
R × B = Deflections in Prime Vertical		+0·10	+0·48	+0·71	+0·73	+0·39	+0·10	−0·10	−0·25	−0·43	−0·34	−0·20	−0·05

Calculated Total Deflection in the Meridian = + 0″·20.

Calculated Total Deflection in the Prime Vertical = + 1 ·14.

LOSALLI.

Height above Mean Sea Level = 1749 feet.

Heights of Compartments in feet.

Radii of Annuli		SECTORS											
		N	N.E.	E.		S.E.	S.		S.W.	W.		N.W.	N.
		$\alpha' = 30°$	$= 60°$	$= 90°$	$= 120°$	$= 150°$	$= 180°$	$= 210°$	$= 240°$	$= 270°$	$= 300°$	$= 330°$	$= 0°$
r'	r_1	$\alpha_1 = 0°$	$= 30°$	$= 60°$	$= 90°$	$= 120°$	$= 150°$	$= 180°$	$= 210°$	$= 240°$	$= 270°$	$= 300°$	$= 330°$
miles 0·25	miles 0·125	1740	1740	1740	1740	1740	1740	1740	1740	1740	1740	1740	1740
0·5	0·25	1740	1740	1730	1730	1730	1730	1730	1730	1730	1720	1740	1740
1	0·5	1730	1730	1730	1730	1730	1730	1720	1720	1720	1700	1730	1730
2	1	1730	1730	1710	1720	1700	1710	1700	1700	1700	1700	1720	1710
4	2	1720	1730	1690	1700	1680	1660	1670	1700	1720	1720	1720	1700
8	4	1730	1700	1750	1720	1700	1680	1680	1730	1750	1720	1700	1720
16	8	1740	1740	1450	1530	1600	1650	1660	1750	1750	1750	1720	1650
32	16	1720	1550	1420	1380	1400	1420	1500	1600	1450	1450	1500	1700
64	32	1620	1600	1500	1600	1700	1500	1500	1500	1360	1450	1350	1700
Sum − 15741 = S		− 271	− 481	−1021	− 891	− 761	− 921	− 841	− 571	− 821	− 791	− 821	− 351
S × ·000817 = R		−0·221	−0·393	−0·834	−0·728	−0·622	−0·752	−0·687	−0·467	−0·671	−0·646	−0·671	−0·287
$\sin \alpha' - \sin \alpha_1 = A$		+0·500	+0·366	+0·134	−0·134	−0·366	−0·500	−0·500	−0·366	−0·134	+0·134	+0·366	+0·500
R × A = Deflections in Meridian		−0·11	−0·14	−0·11	+0·10	+0·23	+0·38	+0·34	+0·17	+0·09	−0·09	−0·25	−0·14
$\cos \alpha' - \cos \alpha_1 = B$		−0·134	−0·366	−0·500	−0·500	−0·366	−0·134	+0·134	+0·366	+0·500	+0·500	+0·366	+0·134
R × B = Deflections in Prime Vertical		+0·03	+0·14	+0·42	+0·36	+0·23	+0·10	−0·09	−0·17	−0·34	−0·32	−0·25	−0·04

Calculated Total Deflection in the Meridian = + 0″·47.

Calculated Total Deflection in the Prime Vertical = + 0 ·07.

TINSIA.

Height above Mean Sea Level = 1776 *feet.*

Heights of Compartments in feet.

Radii of Annuli		SECTORS											
		N.	N.E.	E.		S.E.	S.		S.W.	W.		N.W.	N.
		$\alpha' = 30°$	= 60°	= 90°	= 120°	= 150°	= 180°	= 210°	= 240°	= 270°	= 300°	= 330°	= 0°
r'	r_1	$\alpha_1 = 0°$	= 30°	= 60°	= 90°	= 120°	= 150°	= 180°	= 210°	= 240°	= 270°	= 300°	= 330°
miles 0·25	miles 0·125	1750	1760	1760	1760	1760	1760	1760	1760	1760	1750	1760	1760
0·5	0·25	1720	1740	1740	1740	1740	1750	1750	1760	1740	1720	1740	1740
1	0·5	1700	1740	1740	1720	1720	1740	1740	1740	1700	1680	1740	1700
2	1	1720	1720	1720	1700	1720	1720	1740	1700	1650	1640	1700	1650
4	2	1720	1740	1700	1720	1720	1700	1720	1650	1600	1600	1600	1620
8	4	1700	1720	1750	1750	1700	1750	1650	1500	1500	1450	1450	1550
16	8	1700	1700	1700	1720	1600	1650	1480	1400	1550	1400	1400	1530
32	16	1650	1650	1550	1500	1550	1620	1460	1450	1400	1400	1350	1500
64	32	1500	1500	1350	1350	1350	1600	1500	1450	1400	1300	1150	1350
Sum − 15984 = S		− 824	− 714	− 974	−1024	−1124	− 694	−1184	−1574	−1684	−2044	−2094	−1584
S × ·000817 = R		−0·673	−0·583	−0·796	−0·837	−0·918	−0·567	−0·967	−1·286	−1·376	−1·670	−1·711	−1·294
$\sin\alpha' - \sin\alpha_1 = A$		+0·500	+0·366	+0·134	−0·134	−0·366	−0·500	−0·500	−0·366	−0·134	+0·134	+0·366	+0·500
R × A = Deflections in Meridian		−0·34	−0·21	−0·11	+0·11	+0·34	+0·28	+0·48	+0·47	+0·18	−0·22	−0·63	−0·65
$\cos\alpha' - \cos\alpha_1 = B$		−0·134	−0·366	−0·500	−0·500	−0·366	−0·134	+0·134	+0·366	+0·500	+0·500	+0·366	+0·134
R × B = Deflections in Prime Vertical		+0·09	+0·21	+0·40	+0·42	+0·34	+0·08	−0·13	−0·47	−0·69	−0·84	−0·63	−0·17

Calculated Total Deflection in the Meridian = − 0″·30.
Calculated Total Deflection in the Prime Vertical = − 1 ·39.

SALOT.

Height above Mean Sea Level = 1834 feet.

Heights of Compartments in feet.

Radii of Annuli		SECTORS											
		N.	N.E.	E.		S.E.	S.		S.W.	W.		N.W.	N.
		$\alpha' = 30°$	$= 60°$	$= 90°$	$= 120°$	$= 150°$	$= 180°$	$= 210°$	$= 240°$	$= 270°$	$= 300°$	$= 330°$	$= 0°$
r'	r_1	$\alpha_1 = 0°$	$= 30°$	$= 60°$	$= 90°$	$= 120°$	$= 150°$	$= 180°$	$= 210°$	$= 240°$	$= 270°$	$= 300°$	$= 330°$
miles 0·25	miles 0·125	1820	1820	1820	1820	1820	1820	1770	1770	1820	1820	1820	1820
0·5	0·25	1780	1800	1800	1820	1820	1820	1720	1750	1770	1770	1750	1750
1	0·5	1750	1800	1780	1800	1760	1800	1700	1700	1700	1700	1700	1700
2	1	1750	1750	1740	1750	1720	1740	1650	1650	1600	1600	1600	1650
4	2	1780	1780	1750	1700	1680	1650	1600	1600	1500	1500	1500	1680
8	4	1700	1720	1750	1750	1650	1600	1500	1450	1450	1400	1450	1450
16	8	1700	1740	1700	1730	1700	1650	1500	1450	1450	1450	1400	1450
32	16	1700	1680	1700	1600	1650	1600	1520	1400	1400	1400	1350	1400
64	32	1500	1500	1350	1350	1350	1600	1500	1450	1400	1300	1150	1400
Sum − 16506 = S		−1026	− 916	−1116	−1186	−1356	−1226	−2046	−2286	−2416	−2566	−2786	−2206
S × ·000817 = R		−0·838	−0·748	−0·902	−0·969	−1·108	−1·002	−1·672	−1·868	−1·974	−2·096	−2·276	−1·802
Sin α′ − sin α₁ = A		+0·500	+0·366	+0·134	−0·134	−0·366	−0·500	−0·500	−0·366	−0·134	+0·134	+0·366	+0·500
R × A = Deflections in Meridian		−0·42	−0·27	−0·12	+0·13	+0·41	+0·50	+0·84	+0·68	+0·26	−0·28	−0·83	−0·90
Cos α′ − cos α₁ = B		−0·134	−0·366	−0·500	−0·500	−0·366	−0·134	+0·134	+0·366	+0·500	+0·500	+0·366	+0·134
R × B = Deflections in Prime Vertical		+0·11	+0·27	+0·45	+0·48	+0·41	+0·13	−0·22	−0·68	−0·99	−1·05	−0·83	−0·24

Calculated Total Deflection in the Meridian = 0″·00.

Calculated Total Deflection in the Prime Vertical = − 2 ·16.

KÁMKHERA.

Height above Mean Sea Level* = 1780 *feet.

Heights of Compartments in feet.

Radii of Annuli		SECTORS											
		N.	N.E.	E.		S.E.	S.		S.W.	W.		N.W.	N.
		$\alpha' = 30°$	$= 60°$	$= 90°$	$= 120°$	$= 150°$	$= 180°$	$= 210°$	$= 240°$	$= 270°$	$= 300°$	$= 330°$	$= 0°$
r'	r_1	$\alpha_1 = 0°$	$= 30°$	$= 60°$	$= 90°$	$= 120°$	$= 150°$	$= 180°$	$= 210°$	$= 240°$	$= 270°$	$= 300°$	$= 330°$
miles 0·25	miles 0·125	1770	1770	1770	1770	1760	1750	1750	1770	1760	1770	1770	1770
0·5	0·25	1760	1700	1720	1740	1720	1700	1720	1730	1750	1760	1760	1720
1	0·5	1660	1650	1700	1700	1650	1650	1650	1720	1720	1740	1700	1650
2	1	1650	1600	1600	1600	1550	1550	1700	1700	1720	1700	1650	1640
4	2	1600	1550	1550	1550	1450	1450	1450	1550	1680	1700	1650	1700
8	4	1520	1500	1450	1400	1400	1400	1400	1500	1650	1650	1650	1560
16	8	1480	1460	1380	1350	1350	1350	1400	1500	1600	1700	1700	1550
32	16	1480	1400	1350	1500	1350	1350	1400	1550	1550	1700	1650	1700
64	32	1580	1560	1400	1400	1700	1600	1500	1500	1350	1400	1470	1600
Sum − 16020 = S		−1520	−1830	−2100	−2010	−2090	−2220	−2050	−1500	−1240	− 900	−1020	−1130
S × ·000817 = R		−1·242	−1·495	−1·716	−1·642	−1·708	−1·814	−1·675	−1·226	−1·013	−0·735	−0·833	−0·923
Sin α' − sin α_1 = A		+0·500	+0·366	+0·134	−0·134	−0·366	−0·500	−0·500	−0·366	−0·134	+0·134	+0·366	+0·500
R × A = Deflections in Meridian		−0·62	−0·55	−0·23	+0·22	+0·63	+0·91	+0·84	+0·45	+0·14	−0·10	−0·30	−0·46
Cos α' − cos α_1 = B		−0·134	−0·366	−0·500	−0·500	−0·366	−0·134	+0·134	+0·366	+0·500	+0·500	+0·366	+0·134
R × B = Deflections in Prime Vertical		+0·17	+0·55	+0·86	+0·82	+0·63	+0·24	−0·22	−0·45	−0·51	−0·37	−0·30	−0·12

Calculated Total Deflection in the Meridian = + 0″·93.
Calculated Total Deflection in the Prime Vertical = + 1 ·30.

AHMADPUR.

Height above Mean Sea Level = 1715 feet.

Heights of Compartments in feet.

Radii of Annuli		SECTORS											
		N.	N.E.	E.		S.E.	S.		S.W.	W.		N.W.	N.
		$\alpha' = 30°$	$= 60°$	$= 90°$	$= 120°$	$= 150°$	$= 180°$	$= 210°$	$= 240°$	$= 270°$	$= 300°$	$= 330°$	$= 0°$
r'	r_1	$\alpha_1 = 0°$	$= 30°$	$= 60°$	$= 90°$	$= 120°$	$= 150°$	$= 180°$	$= 210°$	$= 240°$	$= 270°$	$= 300°$	$= 330°$
miles 0·25	miles 0·125	1560	1560	1560	1560	1560	1560	1560	1560	1560	1560	1560	1560
0·5	0·25	1480	1480	1480	1480	1480	1480	1480	1480	1480	1480	1480	1480
1	0·5	1440	1440	1440	1440	1440	1440	1440	1440	1440	1440	1440	1440
2	1	1430	1430	1440	1420	1430	1450	1450	1440	1440	1420	1430	1440
4	2	1440	1450	1430	1430	1400	1420	1450	1450	1460	1450	1440	1470
8	4	1420	1430	1380	1400	1450	1400	1500	1400	1450	1450	1450	1450
16	8	1400	1370	1370	1350	1500	1450	1500	1500	1500	1450	1450	1450
32	16	1420	1370	1400	1450	1450	1600	1450	1600	1550	1500	1600	1550
64	32	1450	1350	1400	1650	1300	1400	1300	1600	1500	1400	1500	1740
Sum − 15435 = S		−2395	−2555	−2535	−2255	−2425	−2235	−2305	−1965	−2055	−2285	−2085	−1855
S × ·000817 = R		−1·957	−2·087	−2·071	−1·842	−1·981	−1·826	−1·883	−1·605	−1·679	−1·867	−1·703	−1·516
$\sin \alpha' - \sin \alpha_1 = A$		+0·500	+0·366	+0·134	−0·134	−0·366	−0·500	−0·500	−0·366	−0·134	+0·134	+0·366	+0·500
R × A = Deflections in Meridian		−0·98	−0·76	−0·28	+0·25	+0·73	+0·91	+0·94	+0·59	+0·22	−0·25	−0·62	−0·76
$\cos \alpha' - \cos \alpha_1 = B$		−0·134	−0·366	−0·500	−0·500	−0·366	−0·134	+0·134	+0·366	+0·500	+0·500	+0·366	+0·134
R × B = Deflections in Prime Vertical		+0·26	+0·76	+1·04	+0·92	+0·73	+0·24	−0·25	−0·59	−0·84	−0·93	−0·62	−0·20

Calculated Total Deflection in the Meridian $= -0''·01$.

Calculated Total Deflection in the Prime Vertical $= +0·52$.

The Group System of Deflections may now be exhibited thus:—

Station	In the Meridian			In the Prime Vertical		
	Deflection as observed	Deflection calculated from the contour of the ground	Residual Deflection due to hidden cause	Deflection as observed	Deflection calculated from the contour of the ground	Residual Deflection due to hidden cause
Daiádhari ...	+1·01 S.	0·39 S.	+0·62 S.	+2·13 W.	0·33 W.	+1·80 W.
Súrantál ...	+0·82 S.	0·11 N.	+0·93 S.	+3·64 W.	1·11 W.	+2·53 W.
N. E. End of Base ...	+1·69 S.	0·07 S.	+1·62 S.	+2·54 W.	1·34 W.	+1·20 W.
Bhaorása ...	+1·17 S.	0·38 S.	+0·79 S.	+0·22 W.	0·76 W.	−0·54 E.
Kaliánpur ...	−0·60 N.	0·20 N.	−0·40 N.	−0·22 E.	1·14 W.	−1·36 E.
Losalli ...	−1·02 N.	0·47 N.	−0·55 N.	−6·38 E.	0·07 W.	−6·45 E.
Tinsia ...	+0·98 S.	0·30 S.	+0·68 S.	...	...	...
Salot ...	...	...	...	−4·49 E.	2·16 E.	−2·33 E.
Kámkhera ...	−2·15 N.	0·93 N.	−1·22 N.	+0·04 W.	1·30 W.	−1·26 E.
Ahmadpur ...	−2·49 N.	0·01 S.	−2·50 N.	+2·27 W.	0·52 W.	+1·75 W.
Mean of the group excluding Kaliánpur	0·00	0·05 N.	0·05 S.	0·00	0·41 W.	0·41 E.

It is interesting to see that a deflection due to the configuration of the ground, of 0″·05 in the meridian, and of 0″·41 in the prime vertical remains uncancelled in the mean of the group.

Dehra Dún:

April 1901.

PART III.

The Pendulum Observations at Kalianpur.

A fixed datum for deflections is unattainable, and it is not possible from deflections alone to determine, whether any particular station is situated vertically over a centre of disturbance. Diagram No. 4 of Chart No. 5 furnishes an instance: if we observed for azimuth at every furlong between Kaliánpur and Losalli, the intervening deflections would probably vary from + 3″, the value at Kaliánpur, to − 3″, the value at Losalli, and would pass through zero. But that the station of no deflection was situated over the centre of the subterranean cause of disturbance would not be a true inference, unless the absolute deflection at Kaliánpur was *proved* to be 3″. Our series of deflections, gradually changing at every furlong, would fit equally well into Diagram No. 3 of Chart No. 5: they would vary then from 0″ to − 6″, instead of from + 3″ to − 3″, and the station of no deflection would be differently situated. Pendulum observations are independent of the initial value of latitude of the reference station, and may possibly be utilized to decide between rival systems of deflections. Pendulum observations were taken by Capt. Basevi at several places between the Himalayas and Cape Comorin: the results at five of his stations, situated on the same meridian, are of value to the present discussion.

Stations on the meridian of 77° 30′	Distance in miles from Kaliánpur	Observed defect in the vibrations of a seconds pendulum in a mean solar day
Usira	193	− 3·60
Pahárgarh	52	− 5·60
Kaliánpur	...	− 3·61
Ahmadpur	35	− 4·38
Badgaon	230	− 4·03

We have to consider, whether Basevi's pendulum results support the "Group" system of deflections or the "Mean of India" system. It is unnecessary here to repeat the Table, which exhibits these two systems, and which is given on page 24*: but as pendulum observations were taken at stations north and south of the area of the group, and astronomical latitudes were also observed, *vide* Tabe following page 14, some results are included in the following Table, which were beyond the scope of the former.

Stations on the meridian of 77° 30′	Distance in miles from Kaliánpur	The "Group" system of deflections in the meridian	The "Mean of India" system of deflections in the meridian
Usira ...	193	6″·03 north	3″·43 north
Kesri ...	112	5″·45 south	8″·05 south
Pahárgarh ...	52	0″·76 north	1″·84 south
Daiádhari ...	35	1″·01 south	3″·61 south
Súrantál ...	8	0″·82 south	3″·42 south
Kaliánpur ...	...	0″·60 north	2″·00 south
Kámkhera ...	9	2″·15 north	0″·45 south
Ahmadpur ...	35	2″·49 north	0″·11 south
Ládi ...	67	5″·34 north	2″·74 north
Takalkhera ...	207	6″·90 north	4″·30 north
Badgaon ...	230	7″·83 north	5″·23 north

The deflection of a plumb-line in a given direction, if due to a subterranean cause, may be either produced by an excess of density situated on the side of the station to which the plumb-line is deflected, or a deficiency of density on the opposite side. In discussing a system of deflections, we have to consider whether a deficiency or an excess is the probable cause.

The "Group" system of Deflections.

(a). *Hypothesis of Excessive Density.* Charts Nos. 1, 2 and 8 shew that almost all the deflections, both in the meridian and prime vertical, can be explained, if we assume the existence of a long dyke of excessive density traversing the group from S. E. to N. W. Its centre line may be imagined to pass some 12 miles south of Bhaorása and almost under the S.W. End of the Base and proceed thence between N.E. End of Base and Kámkhera, between Kaliánpur and Súrantál, between Salot and Daiádhari. The only observed latitude, that is opposed to the

* In this table the deflections have not received the small corrections on account of configuration of surface.

hypothesis of the dyke is that of Tinsia: the only opposing azimuth is that of Ahmadpur, which is a station south of the dyke, and where the plumb-line in the prime vertical might not be affected.*

The pendulum observations show that the intensity of gravity is greater at Kaliánpur than at the two neighbouring stations, and that a maximum value of intensity exists at some point between Ahmadpur and Pahárgarh: it is extremely unlikely that in selecting Kaliánpur itself Basevi should have alighted on the exact position of the maximum, and it is probable that a point will be found where the vibration-number exceeds that of Kaliánpur. The location of this point to the north of Kaliánpur and the discovery there of a very small excess in vertical attraction would confirm the "Group" system: if the point of maximum vertical attraction is found to the south of Kaliánpur, a southerly deflection at Kaliánpur would be expected; this latter, if at all appreciable, would be contrary to the results of the "Group" system: whether it would favor the "Mean of India" system, will be discussed hereafter.†

The extent and influence of the imaginary dyke of excessive density must be gauged from the magnitude of the several deflections: the largest deflection under the "Group" system is that in the prime vertical at Losalli; the inferences are that Losalli must be situated to one side of the dyke in a position, where the horizontal component of the dyke's attraction is a maximum, and that Kaliánpur is nearer the vertical plane passing through the centre of the dyke, and at a point where the horizontal component is small ‡.

(b). *Hypothesis of Deficient Density.* The "Group" system of deflections does not favor any hypothesis of a deficiency of density existing in the vicinity of Kaliánpur. Eight meridional deflections out of nine point inwards, seven deflections in the prime vertical out of eight point inwards, and the pendulum observations show an excess of vertical attraction at the central point of the area of the group.

The "Mean of India" system of deflections.

(a). *Hypothesis of Excessive Density.* No plans, such as Charts Nos. 1, 2 and 3 have been drawn to illustrate the "Mean of India" system, but the results of the system are exhibited in the tables on pages 24 and 39 and in the diagrams of Charts Nos. 4 and 5. The deflections, shown

---

* It is perhaps but an argument in a circle to assume the mean latitude and azimuth of a group correct, and to then locate the cause of individual discrepancies: discrepancies derived from their own mean can only indicate a cause near the centre of the group. But it might happen that the meridional deflections were inwards, and the prime vertical deflections outwards: it might happen that the deflections were all towards a point, at which the pendulum exhibited a deficiency of gravity: if the latitude and the azimuth and the pendulum observations are in accord, the evidence is strong.

† We must not expect future pendulum observations to corroborate either the one system or the other with exactitude. According to the "Group" system the deflection at Kaliánpur is 0″·60 North: but an error of observation exists: the effects of irregularities of subterranean density may not have been completely eliminated in the mean of the group, just as the configuration of the surface was shown in Part II to have a small residual effect: moreover the effects of the Himalayas and of the Ocean, though probably eliminated in the mean of the group, may be different at different stations of the group. If pendulum observations show a slight excess of vertical attraction either north or south of Kaliánpur, they will indicate that no marked attraction exists at Kaliánpur, and the "Group" system will be supported. Similarly the "Mean of India" system does not stand or fall according to whether an exact deflection of 2″ south is proved or not. If any considerable deflection to the south is proved, the "Mean of India" system will be upheld. Broadly speaking, the "Group" system denies the existence of any marked *local* attraction in the meridian at Kaliánpur, whereas the "Mean of India" system imposes on Kaliánpur a marked *local* deflection to the south.

‡ It is unfortunate that the weights of azimuth and latitude observations differ so largely: not only is the weight of an *observed* latitude greater than the weight of an *observed* azimuth, but the errors of *geodetic* latitudes are less than those of *geodetic* azimuths: a value of (O − C) in latitude is thus superior both astronomically and geodetically to the value of (O − C) in azimuth. Moreover a meridional deflection is derived directly from a value of (O − C) in latitude, but a deflection in the prime vertical is obtained by multiplying the value of (O − C) in azimuth by cot. lat. (= 2·2). The weight of a deflection in the meridian is 150 times greater than the weight of a deflection in the prime vertical, even at the distance of Losalli from the station of origin.

in Chart No. 1, require a constant correction throughout of 2″·60 *towards the south*, to bring them into accord with the "Mean of India" system: the deflections shown in Chart No. 2 require a constant correction of 3″ *to the west*. All the deflections in Chart No. 1 will then point to the south: all the deflections in Chart No. 2, with the exception of Losalli and Salot, will point to the west. Under the "Group" system the difficulty of locating the "hidden cause" might be considerable, because the deflections, being mostly small, may possibly be due to variations of density too slight to affect a pendulum. But under the "Mean of India" system, the difficulty of location should be less, as a constant deflection in one direction cannot be imposed on nine stations except by a powerful cause.

If the "Mean of India" system of deflections, as exhibited in the tables on pages 24 and 39, is to be explained on an hypothesis of excessive density, we have to assume the existence of a longitudinal mass lying in the prime vertical south of Kaliánpur. If such a mass were north of Ahmadpur, the deflection at Ahmadpur, which is southerly, would be *towards* the north: the existence of such a mass underlying Ahmadpur itself is contradicted by the pendulum observations there: the deflection at Kámkhera points to no such mass. If then such a mass exists, we can only suppose it to be of small extent and of great density, and situated south and within a mile of Kaliánpur. Such a mass might cause a southerly deflection of 2″ at Kaliánpur and if of compact form, exercise no effect at Kámkhera and Ahmadpur. But if its form were compact, it would not explain the southerly deflections at Tinsia, Losalli, N.E. End of Base and Bhaorása: if its form were elongated its influence would be visible at Kámkhera. A hidden mass of excessive density, situated south of Kaliánpur, and sufficient to produce a southerly deflection there of 2″, might be expected at its summit to show an excess of vertical attraction over that at Kaliánpur, *equivalent to a whole vibration of the second's pendulum per diem*. The vertical attraction at a station, overlying such a mass, would be greater than any value obtained hitherto at inland stations in India from pendulum observations. The probability of any considerable mass of excessive density existing south of Kaliánpur is, in fact, so small, that the "Mean of India" system of deflections depends for its justification mainly on the discovery of a deficiency of density to the north.

(b). *Hypothesis of Deficient Density*. We wish now to see, if we can explain the "Mean of India" system of deflections on an hypothesis of deficient density existing *north* of Kaliánpur. The pendulum observations, *vide* page 38, show that at Pahárgarh the intensity of gravity is remarkably less than at Kaliánpur or Usira, the observed defect in the Vibration-number amounting to 2·00. A point of minimum vertical attraction therefore exists north of Kaliánpur. Under the "Mean of India" system, the deflections at Súrantál, Daiádhari, and Pahárgarh are all *essentially southerly*: it is not possible therefore to locate the deficiency of density, which is now supposed to be deflecting the plumb-line at Kaliánpur to the south, anywhere south of Pahárgarh: we must search for the spot between Pahárgarh and Usira. If we locate the deficiency north of Pahárgarh, it will have to be of large extent if it is to affect the whole Kaliánpur group; we cannot continue our search to an indefinite distance to the north, for the "hidden cause" of the southerly deflection at Kaliánpur must after all be but a *local* cause: the basis of the "Mean of India" system is, that Kaliánpur is peculiarly affected with regard to India, the mean error of India being attributed to this purely local cause. When therefore we reach Pahárgarh, we are approaching the limit allowed by our theory.

Between Pahárgarh and Usira is situated the station of Kesri, and the relative *southerly* attraction here is great, amounting under the "Mean of India" system to 8″. We cannot suppose the existence of a deficiency of density between Pahárgarh and Kesri, because a deficiency, sufficient to repel all the plumb-lines of the group, would deflect the plumb-line at Kesri to the north. We have no alternative but to advance north of Kesri: a sphere, whose radius is 37 miles, the depth of whose centre is 37 miles, whose density is in defect of the surrounding surface density by $\frac{1}{90}$th part of the mean density of the Earth, would, if its highest point was 40 miles north of Kesri, cause a southerly deflection at Kesri of 8″, and at Kaliánpur of 2″: the deflections at intermediate stations would not agree with the results of observation. A long cylinder, lying in the prime vertical 18 miles north of Kesri, whose radius and depth of centre were 18 miles, and whose density was in defect of the surrounding surface density by $\frac{1}{100}$th part of the mean density of the Earth, would cause a northerly deflection at Usira of 3″, and a southerly deflection at Kesri of 7″, at Pahárgarh of 3″, at Daiádhari of 2″·7, at Súrantál of 2″·3, at Kaliánpur of 2″, at Kámkhera of 1″·7, and at Ahmadpur of 0″·8. The effect of such a cylinder on the pendulums at Usira and Pahárgarh would be, that Pahárgarh would exhibit an excess of vertical attraction over Usira. The vertical attraction at Pahárgarh happens to be greatly in defect of that at Usira.

The deflection at Kesri is peculiar and irregular: it must be caused either by a deficiency of matter *north* of Kesri, or by an excess to the *south*. The pendulum observations show an excess of matter *north* of Kesri and a deficiency *south*. The contradiction between the deflection at Kesri and the vibration-number at Pahárgarh leads to the conclusion that the two cannot be due to the same cause, and that therefore the deficiency at Pahárgarh is not sufficient to affect a plumb-line at a distance of 60 miles: if a sufficient deficiency existed at Pahárgarh to deflect the plumb-line at Kaliánpur by 2″, its effect would be visible at Kesri *.

Dehra Dún:

April 1901.

---

* Intermediate latitude stations between Pahárgarh and Kesri, between Kesri and Usira, and between Ladi and Takalkhera are desirable: intervals of 60 and 80 miles are too great. Takalkhera is the station at which Colonel Everest made his celebrated calculation of the attraction of the Mahadeo Pahar range of hills. He attributed the apparent error in the observed latitude at Takalkhera to the presence of this range. But now that latitudes have been observed north and south of Takalkhera, *vide* Table following page 14, it will be seen that he would have been confronted with a similar error, if he had placed the centre of his arc at Ládi or Badgaon.

PART IV.

The influences of the Himalaya Mountains and of the Indian Ocean on the Plumb-line in India.

Forty years ago it was concluded from the data then available, that the attraction of the Himalayas was probably compensated by a deficiency of matter existing below the mountain mass: differences of opinion prevailed as to the manner in which the compensation had arisen, but that there was some counteracting cause, cancelling the attraction of the Himalayas, was generally admitted. The acceptance of this theory ended discussion, and for many years our annual reports have not recognised the possibility of Himalayan attraction affecting a plumb-line in Central India. The data on which the theory was based have been forgotten: the theory itself has remained an unquestioned article of professional faith.

The design of future programmes of astronomical work in India is necessarily influenced by the theories that we have accepted: if those theories are incorrect, we may be working on unprofitable lines of progress. A perusal of our recent reports creates an uneasy feeling that we have been seeking not for information, but for corroboration of accepted conclusions: we have decided to adopt the Clarke spheroid, and we explain contradictory results as being due to subterranean causes. The power of ascribing all anomalies to an invisible cause is a dangerous power, that may be blinding us to true inferences. Now that we have almost come to believe in the existence of an excess of matter, wherever we see a deficiency, and in the existence of a deficiency wherever we see an excess, a periodical reconsideration of data cannot be regarded as superfluous.

The compensation of Himalayan attraction.

The theory that the attraction of the Himalayas is counteracted by an invisible cause is mainly based on the following argument :—*The observed effect of the attraction of the Himalaya Mountains on the plumb-line at Kaliána (in latitude 29° 30′ 48″), the northern terminus of the Indian Arc, is 5″·236. But the attraction of the apparent or superincumbent mass of the Himalayas at that point is sufficient to produce a deflection of 27″·853, as calculated by Archdeacon Pratt of Calcutta**.

There are two assumptions in this argument: it is, firstly, *assumed* that the influence of the Himalayas does not extend to Kaliánpur, and, secondly, that the influence of the Indian Ocean is inappreciable. The deflection of 5″·2 is *assumed* to be the *absolute* deflection at Kaliána†. *It represents in reality the difference of the deflections at Kaliána and Kaliánpur.* It might be inferred from the statement of argument given above, that the calculated deflection at Kaliána due to the Himalayas exceeds the observed deflection by 22″. It is true that Pratt calculated the deflection at Kaliána to be 27″; but he also calculated the deflection at Kaliánpur to be 12″, thus making the *difference* in the deflections at the two places to be 15″. The discrepancy between observed and calculated values is thus 10″ and not 22″. But if we take into account the effects of the Indian Ocean, this discrepancy is further reduced; Pratt calculated that the deflection due to the Indian Ocean at Kaliánpur exceeded that at Kaliána by 3″, and thus the discrepancy between the observed and calculated effects of the Himalayas is reduced to 7″. Pratt calculated the effect of the Indian Ocean, before a single sounding in the deep sea had been taken: he assumed the depth of the Ocean everywhere too small: it is shown hereafter that if correct oceanic depths be employed in the calculation, the discrepancy at Kaliána between calculated and observed values will disappear.

The observed value of the latitude of Dehra Dun is opposed to the theory of Himalayan Compensation.

It is incorrect to regard Dehra Dún as lying at the foot of the great Himalayan mass: it is situated indeed in Sub-Himalayan regions, but the effect of the lower ranges of hills is slight: the plateau of Tibet is the main source of attraction, and its southern scarps are 40 miles from Dehra Dún. In Chart No. 9 is shown a cross section of the Himalayas on the great circle through Dehra Dún in azimuth 30° East of North: this section has been constructed by Colonel St. G. C. Gore, R.E., the Surveyor General of India, whose long experience of these mountains gives great weight to the drawing‡.

Since Pratt made his calculations, latitudes have been observed in Sub-Himalayan regions at Dehra Dún and Mussooree, *vide* Table following page 14. Dehra Dún is 55 miles north of Kaliána§: the deflection at Dehra Dún as observed is apparently 38″ and that at Kaliána 7″; there is thus a decrease of 31″ in 55 miles. *Is such a decrease characteristic of the effect which would be produced by a great mass compensated by underlying deficiencies of matter?*

---

* *Vide* Account of the Principal Triangulation, Ordnance Survey of Great Britain and Ireland, pages 573 and 574. *Vide* also Operations of the Great Trigonometrical Survey of India, Volume V, page XXXI: also Philosophical Magazine, Volume XI, May 1861; also Bull. Acad. Science, St. Petersburg, 1861, tom. iii, pages 396—424, in which the discrepancy between observed and calculated results at Kaliána is called an undisputed fact.

† The observed deflection at Kaliána from modern data is 7″·03.

‡ Mussooree is situated a mile west of Landour and on the same ridge. An idea of the position of Mussooree with reference to the Himalayas can be obtained by supposing it to be identical with the ridge called Landour in Chart No. 9.

§ The relative situations of Dehra Dún, Kaliána, Kaliánpur and other astronomical stations are shown on Chart No. 8.

In view of the enormous dimensions of the Himalayan mass, compared with which the distance of 55 miles separating Dehra Dún and Kaliána is small, I cannot conceive any law of attraction that will satisfy the observed deflections.

If we imagine Himalayan attraction to vary inversely with the distance from a longitudinal axis, a decrease from 38″ to 7″ in 55 miles would place that axis 10 miles north of Dehra Dún: if we imagine Himalayan attraction to vary inversely as the square of the distance, the centre of attraction is placed 36 miles north of Dehra Dún.

The only explanation, that appears acceptable, is that the values 38″ and 7″ do not represent absolute deflections but differential, and that though the *difference* between the deflections at Dehra Dún and Kaliána is doubtless 31″, yet the absolute deflections are 38″ + x, and 7″ + x, where x represents the deflection due to Himalayan attraction at Kaliánpur, the station of origin*.

If we compare the observed deflections, given in the table of latitudes following page 14, at Nojli and Kaliána, we find that the deflection at Nojli is double that at Kaliána. Nojli is 25 miles north of Kaliána, and both are situated on flat low alluvial plains; the *nearest* scarp of the Tibetan plateau is from 90 to 100 miles distant. By what law of attraction is the effect of this distant mass twice as great at Nojli as at Kaliána†?

No hypothetical deficiency of matter below the visible Himalayan Mass will suffice to explain the coexistence of large deflections in Sub-Himalayan regions and no deflections in Central India.

An objection to the accepted theory of Himalayan compensation, is this: subterranean compensation should have a greater absolute effect at stations *near* the Himalayas than at distant stations, whereas the theory seems to imply the opposite. On page 803, Volume 186, Philosophical Transactions of the Royal Society, India's contributions to Geodesy, General Walker writes:—
" It is evident that the effect of the attraction of mountain masses on the plumb-line, which may " be very large in the immediate vicinity of the mountains, will be reduced at a distance in greater " proportion than is assigned by an incomplete application of the law of gravitation, because of the " deficiency in the density of the strata under the mountains, which has not hitherto been allowed " for. Eventually a point must be reached at which the positive attraction of the matter above " will be cancelled by the negative attraction of the deficiency below, and then the mountain masses " will have no influence on the plumb-line".

The effect of a subterranean deficiency of matter varies with the depth to which it extends: if the *same* deficiency of matter be distributed through various depths, the direction of the resultant deflection will be different for each depth, and the deflection will decrease as the depth increases.

Colonel Clarke's formula for the attraction of a mountain mass is

$$\rho h \left(\sin a' - \sin a_1\right) \log_e \frac{r' + \sqrt{r'^2 + h^2}}{r_1 + \sqrt{r_1^2 + h^2}}$$

By taking h negative we can apply this formula to a subterranean deficiency of matter.

* Latitude observations are about to be taken on the meridian of 88° at intervals of 30 miles between Calcutta and Darjeeling. The results will show whether the plumb-lines at stations situated south of Mount Everest and Kinchingunga are deflected to the same extent as at Dehra Dún, and whether the compensation of the eastern Himalayas is more perceptible than that of the western.

† On the meridian of 80°, Rámuápur is 54 miles and Jarúra is 76 miles from the foot of the Sub-Himalayas; the Table following page 14 shows that the deflection at Rámuápur is almost twice as great as at Jarúra.

If the area of the mass as defined by a', a_1 and r', r_1 remains constant, and *if the total deficiency of matter remains constant*, that deficiency may be distributed through any depth h: as h, the depth, increases, ρ the density decreases, and ρh remains constant. It is only in cases, when the depth is so large, that it is necessary to take h^2 into account, that the compensating effect of a deficiency of matter below a mountain mass differs appreciably from the effect of a change in the density of the mountain mass itself.

If we suppose the Himalayas exactly compensated by an underlying cavity, whose form and dimensions are the same as those of the Himalayas inverted, the visible mountain mass will produce northerly deflections at all stations, and the invisible underground cavity will produce deflections similar in amount but southerly: neither at Dehra Dún nor at Kaliána nor at Kalíánpur will there be any deflection perceptible to observation.

Let the imaginary underground cavity be filled with water: then if H be the northerly deflection produced at any station by the visible mountain mass, the southerly deflection at that station due the subterranean matter being water instead of rock, will be $\frac{1 \cdot 5\,\mathrm{H}}{2 \cdot 5}$, the density of water being $= 1$, the density of rock being $= 2 \cdot 5$. The resulting deflections at *all* stations will be as though the visible mountain mass had a density of 1, and the *nearer* stations will be affected by the compensation to a *greater* absolute extent than stations at a distance.

Let H_1, H_2 and H_3 be the *northerly* deflections produced at Dehra Dún, Kaliána and Kalíánpur respectively by the visible superincumbent Himalayan mass; then the *southerly* deflections, at the same stations produced by a deficiency of matter, underlying the Himalayas, equal in amount to that contained in the Himalayas themselves, will be as follows*:—

If the deficiency is distributed through	Dehra Dún	Kaliána	Kalíánpur
A depth of 10 miles	H_1	H_2	H_3
A depth of 100 miles	$\cdot 9\,\mathrm{H}_1$	$\cdot 9\,\mathrm{H}_2$	H_3
A depth of 500 miles	$\cdot 5\,\mathrm{H}_1$	$\cdot 6\,\mathrm{H}_2$	$\cdot 8\,\mathrm{H}_3$
A depth of 1000 miles	$\cdot 3\,\mathrm{H}_1$	$\cdot 4\,\mathrm{H}_2$	$\cdot 6\,\mathrm{H}_3$

* The results in this table have been obtained as follows:—Tables containing the heights of compartments round Dehra Dún, Kaliána and Kalíánpur are given hereafter. In these tables the Himalayan compartments can be identified, as their heights are printed in special type. It is thus easy to find the effect of the Himalayas alone on the plumb-line. In order to find the effect of an equal subjacent mass distributed through a depth D, the depth of the deficiency underlying the Tibetan plateau is taken to be D miles, and as the true height of that plateau is 3 miles, the height of every Himalayan compartment is multiplied by $\frac{\mathrm{D}}{3}$. Thus the expression "distributed through a depth of 1000 miles" is taken to mean that the deficiency underlying *the main plateau* extends to a depth of 330 times the true height of the plateau: the depth of each compartment is then found by multiplying its true height by 330. Thus the depth of the imaginary deficiency is assumed everywhere proportional to the height of the mountains immediately superincumbent. If H_1 is the northerly deflection produced at Dehra Dún by the visible Himalayan mass standing on any compartment, the southerly deflection due to an equivalent deficiency of matter distributed under that compartment through a depth of 1000 miles will be

$$\mathrm{H}_1 \times \left[\frac{\log_e \left\{ r' + \sqrt{r'^2 + (330h)^2} \right\} - \log_e \left\{ r_1 + \sqrt{r_1^2 + (330h)^2} \right\}}{\log_e r' - \log_e r_1} \right]$$

where h is the average height of the Himalayas on the compartment.

Then the ***resultant*** deflections, which will be shown by observations to exist at the three stations, under the combined influence of the attraction of the Himalayas and the repulsion of the subjacent deficiency of matter will be as follows :—

If the deficiency is distributed through	Dehra Dún	Kaliána	Kaliánpur
A depth of 10 miles	0	0	0
A depth of 100 miles	·1 H_1	·1 H_2	0
A depth of 500 miles	·5 H_1	·4 H_2	·2 H_3
A depth of 1000 miles	·7 H_1	·6 H_2	·4 H_3

If the compensating deficiency of matter below the mountain mass be distributed through a depth of 10 miles or less, its effect will exactly counteract the effect of the visible mountain mass, and no deflections will exist at any station.

If the deficiency of matter below the mountain mass be distributed through a depth of 1000 miles, its compensating effect will be greater (*proportionally*) at distant stations than at Dehra Dún, but *will be incomplete at all stations, and the attraction of the Himalayas will then be perceptible to observation at places far distant from them**.

It is not unreasonable to suppose that the mean density of the Himalayan mass and of its underlying strata may be less than the density of surface rocks. But any reduction in the mean density of the Himalayas will affect the deflections at *all* stations proportionally, and no point will then be reached in India, where the influence of the mountain mass will entirely cease. A mere change in the density factor is essentially different to the theory that large deflections may occur in the vicinity of mountains, but that a point must be reached, at which the positive attraction of the matter above will be cancelled by the negative attraction of the deficiency below.

Pratt calculated the deflection due to the visible mountain mass to be at Kaliána 27″ and at Kaliánpur 12″: if we assume the density of the Himalayas to be only one-half of the density of its component rocks, the deflection at Kaliána will be 13″·5 and that at Kaliánpur 6″: no assumed change in the density factor can entirely eliminate *the northerly deflection at Kaliánpur.* Kaliánpur is our reference-station, and the crucial question is: *Is the plumb-line at Kaliánpur affected by Himalayan attraction?* It is difficult to see, how the entire compensation of the attraction at Kaliánpur can co-exist with a large uncompensated effect at Dehra Dún.

No hypothesis of uniform compensation will suffice to explain the contradictory effects of Indian Mountains and Seas.

A comparison of the astronomic and geodetic values of latitude and longitude at stations in South India has brought to light deflections of the plumb-line, which are *apparently outwards*

* If the deficiency of matter be distributed through a depth of 1000 miles, the point "at which the positive "attraction of the matter above will be cancelled by the negative attraction of the deficiency below" will be situated 340 miles south of the Equator. If the deficiency be through a depth of 500 miles, the neutral point will be situated just south of Cape Comorin.

and *towards* the Ocean; these outward deflections are found *at coast stations,* and disappear inland (page 806, Volume 186, Philosophical Transactions of the Royal Society, 1895)*. It has been inferred that the beds underlying the Ocean are of excessive density and *attract* the plumb-lines on the coast. In the case of the Himalayas it is argued, that their effect is compensated at distant stations, but *uncompensated* at near stations: in the case of the Ocean it is argued, that its effect is compensated at distant stations, and *over-compensated* at coast stations. At one moment we are maintaining that Himalayan attraction is *not compensated* at Dehra Dún on account of its proximity to the Himalayas, and at another we are stating that the influence of the Ocean is *more completely* counteracted at Madras than inland on account of its proximity to the Ocean. The submarine excess of density is thus supposed to produce its *maximum* effect at Madras and Mangalore, stations near to the Ocean, whilst the submontane deficiency of density is assumed to produce its *minimum* effect at Dehra Dún and Mussooree, stations near to the mountains. Is there not an inconsistency in these arguments?

All authorities have accepted the theory†, that "under mountains and plains there "is a deficiency of matter approximately equal in amount to the mass above the sea level: "that below Ocean beds there is an excess of matter approximately equal to the deficiency in the "Ocean when compared with rock: *that the amount of matter in any vertical column drawn from "the surface to a level surface below the crust is approximately the same in every part of the Earth.*" (Clarke, Geodesy, page 98). According to the theory of M. Faye the excesses of matter under oceans, and the deficiencies under mountains have been caused by differences of temperature. (Comptes Rendus, Volume XC, page 1185). He points out that at the bottom of the sea at a depth of 4000 metres the temperature is 1° Centigrade, and that at the same depth under a continent the temperature is 149° Centigrade: that the cooling and contraction of the crust are more rapid under seas than under continents. Viewing the problem from the contracted aspect of the Indian data alone, I am not presuming to dispute theories that rest on a world-wide basis: I am merely endeavouring to show that the effects of mountains and seas in India are radically and essentially different. Wherever we observe within sight of mountains we find a deflection (apparently)‡ *towards* those mountains: wherever we observe within sight of seas we find a deflection (apparently)‡ *towards* those seas. (Philosoph. Transact. Royal Society, Volume 186, page 806). At Dehra Dún we find a meridional deflection of 38″ *towards* the Himalayas: let us suppose that instead of a Tibetan plateau there exists a Tibetan sea, whose form, position and dimensions are the same as those of the plateau inverted: the meridional deflection at Dehra Dún due to this hypothetical sea would by analogy be 24″ *towards the south.* But observations at numerous stations on the shores of the Arabian Sea, the Bay of Bengal, and the Indian Ocean lead us to infer that, if a Tibetan sea existed, the deflection at Dehra Dún would be 3″ *towards the north.* At coast stations in South India we find a meridional deflection of 3″ *towards* the Ocean: let us suppose that instead of oceans and seas there exists an equatorial plateau, whose form, position and dimensions are the same as those of the Arabian Sea, the Indian Ocean, and the Bay of Bengal inverted: the meridional deflection at coast stations in South India due to this hypothetical equatorial plateau would by analogy be opposite in sign to what actually now exists: there would in other words be a deflection of 2″ in the opposite direction from the plateau. The analogy is fair that if the Indian Ocean attracts, the equatorial plateau would repel: but actual experience on the mountains of Northern and Central India has taught us, that an equatorial

---

* *Vide* page XXXII, Volume V, Operations of the Great Trigonometrical Survey of India. *Vide* also Philosoph. Magazine, August, 1878. *Vide* also Table following page 14 of this paper.

† Helmert, mathematischen u. physikalischen Theorien der Höheren Geodäsie, Teil II., S. 365.

‡ I say "apparently", because the observed values of deflections are based on an assumed Figure of the Earth and on an assumed absence of attraction at the station of origin. The apparent southerly deflections at the stations of South India, *i.e.*, the southern positive zone, may be due to the unsuitability of the Everest spheroid: but the Longitude Arcs of South India exhibit deflections *towards* the sea, and these cannot well be attributed to errors of spheroid. A *meridional* deflection towards the land is exhibited at Bombay and Waltair, but the deflection in the Prime Vertical at these places is towards the sea: and the *resultant* deflection of the plumb-line is therefore sea-wards. The statement that wherever we observe within sight of the sea we find a sea-ward deflection, is therefore correct.

plateau of the dimensions of the Indian Ocean inverted would assuredly produce large deflections *towards* itself at stations on its perimeter.

The discordance between the effects of mountains and seas cannot be attributed to the presence of water in the latter. We can imagine the water compressed in volume, until its density equals that of rock: it will then occupy $\frac{2}{5}$ths of the volume of the Ocean, and a vacuity equal to $\frac{3}{5}$ths of the Ocean will exist. The presence of water does not alter the character of deflections, though it modifies their amounts. It may be pointed out that the theory of compensation as quoted above, expressly states that the deficiency of matter under mountains is equal to the mass of those mountains, but that the excess of matter under oceans is equal to the *deficiency in the ocean, when compared with rock*: and that *the amount of matter in any vertical column is approximately the same.*

Necessity for re-calculation.

When writing Part I of this paper, I compared the Himalayan heights and Oceanic depths assumed by Pratt with those that have been given by recent explorations and surveys: in making the comparison my object was to learn, if modern geographical knowledge would justify a decrease in Pratt's calculations of the attraction, the magnitude of his results having always been considered an obstacle to their acceptance.

I did not then see the significance of Pratt having taken his heights and depths *too small.* A chart of the Ocean, Chart No. 8, in which the depths at several places are marked, is attached to this paper: at each place the true depth derived from modern Admiralty charts is given in roman figures with an *a* attached to them, and the depth assumed by Pratt is given in italic numbers with a *p* attached to them.

The errors in Pratt's assumed depths will be seen to be so great that his calculated values for the negative attraction of the Ocean must be rejected*. It is useless to discuss discrepancies between calculated and observed results, unless the calculated results are based on correct data. It is impossible to examine the question, unless a re-calculation is undertaken.

The method of Calculation.

The method of calculation, that has been adopted, is that given by Colonel Clarke in his work on Geodesy. Round each station, as a common centre, sixteen circles have been drawn on the maps, and through each station a series of thirty-six radial lines: the country round each station has been thus divided into a series of four-sided compartments: let a_1 and a' be the azimuths of two consecutive lines, and r_1 and r' the radii of two consecutive circles; then Colonel Clarke shows that the deflection in the direction north caused by the mass of the compartment contained between limits a_1 and a', and r_1 and r' is

$$12''\cdot 44 \frac{\delta}{\Delta} h (\sin a' - \sin a_1) \log_e \frac{r'}{r_1},$$

where δ is the density of the mass, Δ the mean density of the earth, and h the average height of the upper surface of the mass above the station.

---

* Pratt assumed the depths at certain points in the Arabian Sea, Bay of Bengal and Indian Ocean, and then imagined the floor of the Ocean to slope up gradually from these points to the coast lines.

The approximate deflection in the prime vertical was derived from the formula

$$12''\cdot 44 \frac{\delta}{\Delta} h (\cos a' - \cos a_1) \log_e \frac{r'}{r_1}.$$

The radius r' was taken equal to $2r_1$, and thence $\log_e \frac{r'}{r_1}$ is equal to 0·693. The radial lines were drawn at equal intervals of 10° in azimuth.

The calculated values of deflections vary directly with the value adopted for the ratio of the surface density to the mean density of the earth: it is therefore easy after the calculation to find the effect of a change in this ratio. The ratio of $\frac{1}{2}$, adopted in the calculations on page 25, is probably too large, and was selected as being a round number. Colonel Clarke has shown in his Geodesy that the ratio of the surface density to the mean density of the earth can be expressed in terms of the ellipticity of the surface, and as the spheroid of reference, on which our triangulation has been projected, has an ellipticity of $\frac{1}{300\cdot8}$, I have, in order to maintain consistency, taken the ratio of the surface density to the mean density to be $\frac{1}{2\cdot2}$.

The formula for the deflection in the meridian for each sector thus becomes

$$12''\cdot 44 \times \frac{1}{2\cdot 2} \times 0\cdot 693 \times \frac{[h] - 15\text{H}}{5280} \times (\sin a' - \sin a_1)$$

$$= 0''\cdot 000742 \{[h] - 15\ \text{H}\} (\sin a' - \sin a_1),$$

where H = the height of the station, and $[h]$ = the sum of the average heights of the fifteen compartments in a sector.

If commencing due west of a station and proceeding clockwise through north, east and south, we number the sectors 1, 2, 3 to 36 then the factor $0''\cdot 000742\ (\sin a' - \sin a_1)$ for each sector may be computed and tabulated: the following values of factors for the several sectors have been employed, taking f_1 = factor for meridional deflections and f_2 = factor for deflections in the prime vertical:—

TABLE OF FACTORS.

	SECTORS								
	1,18,19,36	2,17,20,35	3,16,21,34	4,15,22,33	5,14,23,32	6,13,24,31	7,12,25,30	8,11,26,29	9,10,27,28
$\sin a' - \sin a_1$	·015	·045	·074	·100	·123	·143	·158	·168	·174
$\cos a' - \cos a_1$	·174	·168	·158	·143	·123	·100	·074	·045	·015
$\cdot 000742 (\sin a' - \sin a_1) = f_1$	·000011	·000033	·000055	·000074	·000091	·000106	·000117	·000125	·000129
$\cdot 000742 (\cos a' - \cos a_1) = f_2$	·000129	·000125	·000117	·000106	·000091	·000074	·000055	·000033	·000011

The factors were made positive; as it was desirable that meridional deflections to the south should be positive, meridional deflections have been given the same sign as $[h] - 15$ H for southern sectors, and the opposite sign for northern sectors: deflections in the prime vertical have been given the same sign as $[h] - 15$ H for western sectors, and the opposite sign for eastern sectors.

Account had to be taken of the presence of sea-water in the Ocean: supposing the mean density of the earth to be 5·7* and the surface density of India to be 2·6, then the *difference* between the density of the Ocean and the surface density of India was $2 \cdot 6 - 1 = 1 \cdot 6$: therefore in the formula the density factor for depths was $\frac{1 \cdot 6}{5 \cdot 7} = \frac{3}{5} \times \frac{1}{2 \cdot 2}$ (nearly) $= \frac{3}{5}$ths of the density factor for heights. In order to preserve one formula for both heights and depths, the latter have been multiplied by $\frac{3}{5}$: *the depths entered in the following tables are consequently but three-fifths of the depths actually shown on Admiralty charts.*

The calculation has been extended to a distance of 4000 miles from each station: it was necessary to extend the calculation to such a distance, that irregular masses beyond might be presumed to affect all stations in India similarly: the average elevations of all continental regions north of India are fairly well known, and the depth of the Ocean for many thousand miles south of India has been ascertained. By limiting the calculation to 4000 miles it is probable that the resulting meridional deflection at Punnæ, the southernmost station, is slightly too large—say by 1″—relatively to the deflection at Mussooree. Mussooree is 1500 miles north of Punnæ, and consequently a strip of the southern Ocean 1500 miles in width, that did not enter into the calculation for Mussooree, has been included in that for Punnæ.

Beyond 4000 miles the correction for curvature rapidly increases with distance, and tends to lessen the effects of mountains and seas. The heights and depths of compartments situated at distances of 500 miles and more from the station of calculation have been corrected for curvature.

Annulus		$\text{Log}_e \frac{r'}{r_1} = A$	$\text{Log}_e \frac{\tan \frac{1}{2} r'}{\tan \frac{1}{2} r_1} + \cos \frac{1}{2} r' - \cos \frac{1}{2} r_1 = B$	Factor for Curvature $= \frac{B}{A}$
r_1 in miles	r' in miles			
500	1000	0·693	0·689	0·994
1000	2000	„	0·675	0·974
2000	4000	„	0·617	0·890
4000	8000	„	0·424	0·612
8000	16000	„	0·093	0·134

When a compartment includes both land and sea, the mean height or depth is taken to be $H \times \frac{L}{L+S} - \frac{3D}{5} \times \frac{S}{L+S}$, where L = area and H = mean height of the land portion, and S = area and D = mean depth of the sea portion.

* Helmert, mathem. u. physikal. Theorien der Höheren Geodäsie, Teil II, S. 392,500.

An error of 1,000 feet in the adopted height of a compartment will, if the compartment be situated due north or south, produce an error of 0″·13 in the resulting deflection. Nowhere within the limits of India is a compartment likely to be in error by 1,000 feet; it is possible that Himalayan compartments may occasionally be in error by as much as 2,000 or 3,000 feet; if ten such compartments have a probable error of 3,000 feet, the probable error of the resulting deflection will be about 1″.

In determining the heights of Tibet, I have made use of the explorations of Prejevalsky, Pevtsof, Littledale, Bonvalot, Sven Hedin and Deasy. The depths of the sea have been taken from Admiralty charts, and Captain Heming, R.N., the Superintendent of the Marine Survey of India, has supplied me with valuable information. I made much use of Captain Basevi's heights of compartments in the vicinity of Mussooree.

Selection of Stations for Calculation.

Chart No. 8, illustrates the positions of the 12 stations selected for calculation. The first stations chosen were Kaliána, Kaliánpur and Dámargída, being Pratt's three stations: to these were added the stations of Dehra Dún and Mussooree in the north and of Punnæ in the extreme south of India: there were then six stations situated on the central meridian of India.

It was intended to end the calculation with these six stations, but when the result at Punnæ alone was found to be at serious variance with the result of observations, it was considered advisable to include more coast stations, and the four stations of Bombay, Mangalore, Madras and Waltair were added (*vide* Chart No. 8).

The station of Calcutta was added, because observations give it a *southerly* deflection with reference to the deflection at Kaliánpur, and it was difficult to conceive, how the deflection at a station situated like Calcutta due south of the Himalayas and due north of the Bay of Bengal could be less northerly than at Kaliánpur, which is south-west of the Himalayas.

Test Station. If the theory of Himalayan compensation is correct, and if no northerly attraction exist at Kaliánpur, the values of (O – C) in latitude will not be large at stations remote from India and will show no persistence in sign. If on the other hand the plumb-line at Kaliánpur suffers a northerly deflection, such as its situation would lead one to expect, then the value of (O – C) at stations removed from the influences of the Himalayas and Ocean should be persistently positive: Mandalay would be a valuable latitude station, but the deflection of its plumb-line cannot be calculated, because the heights of the mountains north of Burma are unknown: these mountains rise to great elevations, and the mean heights of areas in this region could not be estimated within 3,000 or 4,000 feet. The impossibility of calculating the deflection at Mandalay left Baluchistan the most suitable place for a test station: the longitude station at Quetta* would have served the purpose, had it not been too close to hills. Another station was therefore selected in a flat open valley with no hills in the immediate vicinity. The station has been named "Beyond-Quetta", and the deflection of its plumb-line has been calculated. A northerly deflection at Beyond-Quetta is to be expected: on the north side is the Afghan plateau, and on the south the low Jacobabad desert and the Arabian Sea, the matter to the north being in visible excess of that to the south. Therefore, if there is no northerly attraction at Kaliánpur, the value of (O – C) at Beyond-Quetta should be negative. If, on the other hand, there is a large northerly deflection at Kaliánpur, it will exceed in amount the northerly deflection at Beyond-Quetta, and the value of (O – C) at the latter will be positive. Beyond-Quetta is thus a test station: if it furnishes a negative value of (O – C) the theory of Himalayan compensation

* No latitude observations have been taken in Baluchistan.

will be confirmed: if it furnishes a positive value of (O−C), the attraction of the Himalayas at Kaliánpur will be shown to be probably appreciable.

Tests of the Formulæ.

Being uncertain as to the most favourable forms and dimensions to give to compartments, I calculated the deflection at Kaliánpur under different conditions: *firstly*, the country round Kaliánpur to a distance of 4000 miles was intersected by radial lines at 10° interval in azimuth and by circles, whose radii increased in geometrical progression with a factor of 2: the deflection under these conditions was calculated as 37″·6. *Secondly*, the same area was intersected by radial lines at 15° interval in azimuth and by circles, whose radii increased in geometrical progression with a factor of $\frac{3}{2}$: the deflection under these conditions was calculated as 36″·4. The discrepancy between the two values of the deflection derived from different systems of dissection was 1″·2.

Mussooree and Dehra Dún are within 8 miles of each other, and observations show that the deflection of the plumb-line at the two places is almost identical: the deflection derived from the following calculations is at Mussooree 73″·5, and at Dehra Dún 73″·2. A discrepancy of only 0″·3 was reassuring, in that both Mussooree and Dehra Dún are situated in hilly regions, the heights of whose compartments are difficult to average, and whereas Dehra Dún lies in an open valley with no hills within 4 miles, Mussooree stands at the summit of a precipitous hill*.

As Rájpur, situated only 5 miles from Dehra Dún, shows a deflection exceeding that at Dehra Dún by 10″, *vide* Table following page 14, a rough calculation of its deflection was made to test the practical application of the formula: the hills rise at 700 yards from Rájpur: if two circles be drawn one with a radius of 700 yards and one with a radius of 5 miles, and four radial lines one in azimuth 60° west of north, another in azimuth 30° east of north, a third in azimuth 30°

---

* The error arising from the adoption of a ratio $\frac{r'}{r_1} = 2$ may be found: we will take an extreme case and suppose that the height of one half of a compartment is 10000 feet and the height of the other half 0. By our system of averaging we take the height of the whole compartment to be 5000 feet. The deflection, due to the compartment, as calculated, is 0″·000186 × 5000 $\log_e 2$ = 0″·64: the true deflection, if the inner half of the compartment is 10000 feet high, is 0″·000186 × 10000 $\log_e \sqrt{2\cdot5}$ = 0″·85: the true deflection, if the outer half of the compartment is 10000 feet high, is 0″·000186 × 10000 $\log_e \frac{2}{\sqrt{2\cdot5}}$ = 0″·44. The error in the deflection due to this compartment arising from the adoption of the ratio $\frac{r'}{r_1} = 2$, is 0″·20.

In such an extreme case an error *must* obtain, whatever value of $\frac{r'}{r_1}$ be adopted. If we had taken $\frac{r'}{r_1} = \frac{3}{2}$, instead of 2, then the deflection due to the compartment as calculated would have been 0″·38. The true deflection, if the inner half of the compartment had been 10000 feet high, would have been 0″·45. The true deflection, if the outer half of the compartment had been 10000 feet high, would have been 0″·30. The error with a ratio of $\frac{r'}{r_1} = \frac{3}{2}$ is 0″·08.

On the other hand the smaller the ratio, the greater the number of annuli requisite, and, if e be the probable error of the deflection dne to one compartment, the probable error of the deflection due to a sector is $e\sqrt{n}$, where n = nnmber of annuli.

The *magnitude* of distant compartments is not decreased by decreasing the ratio of $\frac{r'}{r_1}$: when $\frac{r'}{r_1} = 2$, the radial length of a compartment becomes 1000 miles, when $r_1 = 1000$ miles: when $\frac{r'}{r_1} = \frac{3}{2}$, the radial length becomes 1000 miles when $r_1 = 2000$ miles: a decrease in the ratio postpones but does not eliminate the necessity of large compartments. The question of the ratio was anxiously considered, but I unfortunately overlooked the possibility of altering the ratio at different distances from the station.

east of south, and a fourth 60° west of south, and if the average height of the enclosed area to the north be taken as 1,500 feet above Rájpur and the average height of the valley to the south as 800 feet below Rájpur, the deflection at Rájpur due to the hills and valleys, that are situated within a radius of 5 miles, will be

$$12''\cdot 44 \times \frac{2\cdot 6}{5\cdot 7} \times \frac{1500 + 800}{5280} \times (\sin 60^\circ + \sin 30^\circ) \log_e 12\cdot 5 = 9''.$$

It is but natural that the excess of the deflection at Rájpur over that at Dehra Dún, as calculated, should be *less* than the observed value, seeing that we have omitted from the calculation any consideration of the fact that Rájpur is 5 miles nearer than Dehra Dún to the Tibetan plateau.

TABLES

OF

HEIGHTS OF COMPARTMENTS.

Himalayan Heights are shown in Roman Figures, thus,					1769.
Continental Heights	,,	in Ordinary	,,	,,	1769.
Oceanic Depths	,,	in Italic	,,	,,	*1769.*

To allow for the presence of sea-water Oceanic depths as entered in the tables are $\frac{3}{5}$ths of the true depths derived from Admiralty charts.

When every compartment of an annulus has the same average height, that height is entered in the two centre columns only.

Height above Mean Sea Level = 6920 feet.

Heights of Compartments in feet and Calculation of Deflection of Plumb-line. North Side.

Radii of Annuli		SECTORS																	
		W.				N.W.				N.					N.E.				E.
		a' = 280°	290°	300°	310°	320°	330°	340°	350°	0°	10°	20°	30°	40°	50°	60°	70°	80°	90°
r′	r_1	a_1 = 270°	280°	290°	300°	310°	320°	330°	340°	350°	0°	10°	20°	30°	40°	50°	60°	70°	80°
Miles 0·125	Miles 0	6840	6850	6850	6850	6860	6870	6880	6890	6900	6890	6880	6880	6870	6870	6860	6860	6850	6860
0·25	0·125	6400	6400	6400	6400	6420	6440	6460	6480	6500	6600	6570	6530	6500	6410	6400	6600	6700	6800
0·5	0·25	6450	6300	6160	6070	5980	5920	5890	5900	5910	6050	6150	6100	6040	6150	6300	6320	6380	6450
1	0·5	6680	6620	6640	6500	6280	6080	5890	5740	5560	5510	5500	5500	5590	5550	5700	5780	5900	6100
2	1	6500	5900	5900	6300	6000	5500	4900	4400	4100	3900	4000	4600	5200	5500	5900	6400	6650	6700
4	2	6200	5900	5600	5300	4700	4100	3300	3700	4000	4300	4300	3800	3800	4000	4400	5000	5700	6200
8	4	5000	4900	4300	3700	3300	3400	3900	4200	4900	4990	5000	5300	5000	4800	4600	4500	4800	5700
16	8	2200	2500	2600	3800	4500	4000	3200	3200	3600	4100	5000	6000	6500	5500	5000	5000	5000	5000
32	16	1700	2800	3000	3600	3600	5000	5300	5800	5300	4500	4600	5700	5800	4500	4500	5300	3800	3500
64	32	1700	3400	4500	5500	5300	5500	6000	7000	8500	9500	10000	12500	13500	9500	11000	12000	7500	6500
128	64	800	1000	1400	2800	3500	3800	5000	11500	11500	11000	12000	13000	14800	13500	11500	11500	10000	7000
256	128	650	650	700	900	1500	2000	9000	12000	12000	12000	13000	14000	14000	14000	14000	15000	14000	10000
512	256	600	600	2000	2000	5000	7000	7000	8000	9000	10000	9000	12000	14000	16000	15000	14000	15000	13000
1024	512	2000	4000	5400	3000	3000	3500	5000	4000	3000	3000	3000	3000	4000	5000	8000	11000	11000	9000
2048	1024	2000	2000	600	300	600	600	600	600	600	600	1000	4000	4000	3000	3000	3000	3000	2000
Beyond		1000	0	0	0	0	0	0	0	0	0	0	0	200	200	0	9500	8000	8000
Sum − 103800		− 53920	− 50830	− 48600	− 47630	− 44120	− 40960	− 32360	− 21280	− 19330	− 17840	− 14680	− 1770	+ 5130	− 190	+ 1500	− 1900	− 6370	− 17850
Correction for Curvature		− 200	− 100	0	0	0	0	0	0	0	0	0	− 100	− 100	− 100	− 100	+ 900	+ 700	+ 800
Sum = A		− 54120	− 50930	− 48600	− 47630	− 44120	− 40960	− 32360	− 21280	− 19330	− 17840	− 14680	− 1870	+ 5030	− 290	+ 1400	− 1000	− 5670	− 17050
$f_1 \times A$		− 0·6	− 1·7	− 2·7	− 3·5	− 4·0	− 4·3	− 3·8	− 2·7	− 2·5	− 2·3	− 1·8	− 0·2	+ 0·5	0·0	+ 0·1	− 0·1	− 0·2	− 0·2
$f_2 \times A$		−7·0	−6·4	−5·7	−5·0	−4·0	−3·0	−1·8	−0·7	−0·2	−0·2	−0·5	−0·1	+0·4	0·0	+0·1	−0·1	−0·7	−2·2

MUSSOOREE.

Height above Mean Sea Level = 6920 feet.

Heights of Compartments in feet and Calculation of Deflection of Plumb-line. **South Side.**

SECTORS																		Radii of Annuli	
E.				S.E.				S.					S.W.				W.		
α' = 100°	110°	120°	130°	140°	150°	160°	170°	180°	190°	200°	210°	220°	230°	240°	250°	260°	270°		
α_1 = 90°	100°	110°	120°	130°	140°	150°	160°	170°	180°	190°	200°	210°	220°	230°	240°	250°	260°	r'	r_1
6850	6860	6850	6850	6850	6820	6820	6810	6800	6810	6810	6820	6810	6790	6800	6800	6800	6820	Miles 0·125	Miles 0
6800	6830	6850	6800	6700	6500	6350	6350	6350	6400	6500	6530	6560	6580	6610	6590	6560	6530	0·25	0·125
6510	6600	6750	6720	6600	6440	6250	6070	5900	5910	6010	6120	6170	6200	6260	6360	6490	6560	0·5	0·25
6300	6350	6500	6550	6360	6100	5850	5600	5500	5550	5600	5800	6050	6250	6350	6500	6500	6700	1	0·5
6800	6500	6000	5800	5600	5500	5200	4900	4700	4800	5200	5300	5200	4980	4990	5300	6000	6680	2	1
6700	6600	5800	5300	5000	4400	4500	4500	4200	3500	3600	3800	3700	3900	4300	4800	5600	6300	4	2
6700	6600	5200	4400	3900	3500	3100	2900	2800	2800	2800	2800	2900	3000	3300	3500	4000	4500	8	4
5000	6500	4300	3600	3100	2700	2400	2000	2000	2100	2100	2100	2000	1900	1900	1900	2000	2000	16	8
4000	4000	4000	4000	3000	2200	1300	1400	1700	1600	1500	1500	1600	1600	1600	1500	1700	1700	32	16
4000	3800	3400	3800	2800	2600	1800	900	830	830	800	800	800	800	800	800	850	900	64	32
6000	3000	2000	2900	1500	1000	800	800	800	750	750	750	750	750	750	750	750	750	128	64
9000	4000	1000	1200	550	550	550	550	550	600	900	900	850	800	750	750	650	650	256	128
10000	3000	400	300	300	300	600	900	900	1300	1200	1200	1200	900	700	700	600	600	512	256
6000	1000	100	900	100	600	1100	1300	1400	1400	1800	1000	200	0	0	200	1000	500	1024	512
1000	1000	100	0	1000	3500	6000	6000	5000	2000	2500	6000	6500	6300	4500	600	1000	1000	2048	1024
9500	8000	2000	0	0	0	9500	8600	8000	6500	5000	5000	6000	2000	0	1000	1000	1000	Beyond	
− 28490	− 46020	− 53400	− 51530	− 59290	− 64910	− 79500	− 80230	− 79170	− 74760	− 72540	− 76200	− 78320	− 74440	− 69990	− 63750	− 59100	− 57430	Sum − 103800	
+ 1000	+ 800	+ 200	0	0	+ 100	+ 1200	+ 1100	+ 1000	+ 800	+ 600	+ 700	+ 800	+ 400	+ 100	− 100	− 100	− 100	Correction for Curvature	
− 27490	− 45220	− 53200	− 51530	− 59290	− 64810	− 78300	− 79130	− 78170	− 73960	− 71940	− 75500	− 77520	− 74040	− 69890	− 63850	− 59200	− 57530	Sum = A	
− 0·3	− 1·5	− 2·9	− 3·8	− 5·4	− 6·9	− 9·2	− 9·9	− 10·1	− 9·5	− 9·0	− 8·8	− 8·2	− 6·7	− 5·2	− 3·5	− 2·0	− 0·6	$f_1 \times A$	
−3·5	−5·7	−6·2	−5·5	−5·4	−4·8	−4·3	−2·6	−0·9	−0·8	−2·4	−4·2	−5·7	−6·7	−7·4	−7·5	−7·4	−7·4	$f_2 \times A$	

Deflection in Meridian = S − N = − 73″·5
Do. Prime Vertical = W − E = − 41 ·1

DEHRA DUN.

Height above Mean Sea Level = 2239 feet.

Heights of Compartments in feet and Calculation of Deflection of Plumb-line. **North Side.**

Radii of Annuli		SECTORS																	
		W.				N.W.				N.				N.E.					E.
		a' = 280°	290°	300°	310°	320°	330°	340°	350°	0°	10°	20°	30°	40°	50°	60°	70°	80°	90°
r'	r_1	a_1 = 270°	280°	290°	300°	310°	320°	330°	340°	350°	0°	10°	20°	30°	40°	50°	60°	70°	80°
Miles 0·125	Miles 0									2240	2240								
0·25	0·125									2240	2240								
0·5	0·25									2250	2250								
1	0·5	2200	2200	2220	2220	2240	2240	2250	2300	2300	2300	2300	2300	2300	2300	2300	2280	2280	2260
2	1	2200	2150	2210	2240	2300	2350	2500	2600	2600	2380	2500	2500	2500	2500	2600	2500	2400	2300
4	2	2050	2050	2200	2270	2400	2500	2700	3000	3000	3000	3000	3000	3000	3000	2800	3000	2700	2500
8	4	1920	2000	2200	2300	2500	3000	3300	3700	3800	3500	3400	3300	3400	3600	3800	4000	3900	3900
16	8	1800	1500	1600	2000	2500	3500	4000	4000	4500	4300	4200	4200	4800	5000	4700	4700	5000	4800
32	16	1800	1600	1500	2500	3000	3600	4000	4800	4500	4000	4000	4100	4200	4300	4100	3900	3700	3500
64	32	1500	3000	3500	4000	5000	5500	6000	7000	7500	7600	8000	9000	11000	12000	9000	8000	6000	4000
128	64	900	1000	2000	3000	3600	5500	6000	9000	10000	11000	12000	13000	15000	13000	11500	11500	10000	7000
256	128	650	650	700	900	1500	2000	9000	12000	12000	12000	13000	14000	14000	14000	14000	15000	14000	10000
512	256	600	600	2000	2000	5000	7000	7000	8000	9000	10000	9000	12000	14000	16000	15000	14000	15000	13000
1024	512	2000	4000	5400	3000	3000	3500	5000	4000	3000	3000	3000	3000	4000	5000	8000	11000	11000	9000
2048	1024	2000	2000	600	300	600	600	600	600	600	600	1000	4000	4000	3000	3000	3000	3000	2000
Beyond		1000	0	0	0	0	0	0	0	0	0	0	0	200	200	0	*9500*	*8000*	*8000*
Sum − 33585		− 8475	− 6345	− 2965	− 2365	+ 4545	+ 12195	+ 23255	+ 31905	+ 33705	+ 34585	+ 36305	+ 45305	+ 53305	+ 54805	+ 51705	+ 44285	+ 41885	+ 27165
Correction for Curvature		− 200	− 100	0	0	0	0	0	0	0	0	0	− 100	− 100	− 100	− 100	+ 900	+ 700	+ 800
Sum = A		− 8675	− 6445	− 2965	− 2365	+ 4545	+ 12195	+ 23255	+ 31905	+ 33705	+ 34585	+ 36305	+ 45205	+ 53205	+ 54705	+ 51605	+ 45185	+ 42585	+ 27965
$f_1 \times A$		−0·1	−0·2	−0·2	−0·2	+0·4	+1·3	+2·7	+4·0	+4·3	+4·5	+4·5	+5·3	+5·6	+5·0	+3·8	+2·5	+1·4	+0·3
$f_2 \times A$		−1·1	−0·8	−0·3	−0·3	+0·4	+0·9	+1·3	+1·1	+0·4	+0·4	+1·2	+2·5	+3·9	+5·0	+5·5	+5·3	+5·3	+3·6

The station at Dehra Dún selected for calculation is the Zenith Sector Pillar in the Haig Observatory in the present office of the Trigonometrical Survey. Its latitude was observed by Captain Lenox Conyngham in November 1899. It is situated 2862 feet south of the old latitude station of Dehra Dún.

At Dehra Dún	Observed Latitude = O	Geodetic Latitude = C	O − C
Old Station ...	30° 19′ 19″·56	30° 19′ 57″·38	− 37″·82
New Station ...	30 18 51 ·92	30 19 29 ·04	− 37 ·12

DEHRA DUN.

Height above Mean Sea Level = 2239 *feet.*

Heights of Compartments in feet and Calculation of Deflection of Plumb-line. **South Side.**

SECTORS																		Radii of Annuli	
E.				S.E.				S.				S.W.					W.		
$a' =$																			
100°	110°	120°	130°	140°	150°	160°	170°	180°	190°	200°	210°	220°	230°	240°	250°	260°	270°		
$a_1 =$																			
90°	100°	110°	120°	130°	140°	150°	160°	170°	180°	190°	200°	210°	220°	230°	240°	250°	260°	r'	r_1
																		Miles	Miles
								2230	2230									0·125	0
								2200	2200									0·25	0·125
								2190	2190									0·5	0·25
2240	2230	2220	2210	2200	2190	2180	2170	2160	2160	2160	2160	2170	2180	2180	2190	2200	2200	1	0·5
2400	2400	2400	2100	2100	2100	2000	2000	2050	2080	2100	2100	2100	2100	2100	2140	2150	2100	2	1
2400	2400	2400	2100	2100	2050	2000	2000	2000	2000	2100	2000	2000	2000	2000	2020	2050	2000	4	2
3800	3900	3600	2000	2000	2000	2200	2200	2200	2200	2150	2100	2050	2000	2000	1950	1950	1900	8	4
4000	3600	2800	2200	1900	1700	1500	1400	1700	2200	2400	2300	2200	2200	2400	2600	2300	2000	16	8
3500	3300	3000	2500	2000	1500	1200	1300	1500	1300	1200	1200	1100	1100	1200	1450	1600	1700	32	16
4000	3000	3000	4000	3000	2300	1100	1000	850	850	850	850	850	900	900	900	950	1100	64	32
6000	3000	2000	4000	2000	800	700	700	700	700	750	750	750	800	800	850	900	900	128	64
9000	4000	1000	1200	550	550	550	550	550	600	900	900	850	800	750	750	650	650	256	128
10000	3000	400	300	300	300	600	900	900	1300	1200	1200	1200	900	700	700	600	600	512	256
6000	1000	100	900	100	600	1100	1300	1400	1400	1800	1000	200	0	0	200	1000	500	1024	512
1000	1000	100	0	*1000*	*3500*	*6000*	*6000*	*5000*	*2000*	*2500*	*6000*	*6500*	*6300*	*4500*	*600*	1000	1000	2048	1024
9500	*8000*	*2000*	*0*	*0*	*0*	*9500*	*8600*	*8000*	*6500*	*5000*	*5000*	*6000*	*2000*	0	1000	1000	1000	Beyond	
+ 15645	− 4365	− 8175	− 5685	− 11945	− 16605	− 29565	− 28275	− 26185	− 20905	− 19085	− 23635	− 26225	− 22515	− 18665	− 13045	− 10845	− 11545	Sum − 33585	
+ 1000	+ 800	+ 200	0	0	+ 100	+ 1200	+ 1100	+ 1000	+ 800	+ 600	+ 700	+ 800	+ 400	+ 100	− 100	− 100	− 100	Correction for Curvature	
+ 16645	− 3565	− 7975	− 5685	− 11945	− 16505	− 28365	− 27175	− 25185	− 20105	− 18485	− 22935	− 25425	− 22115	− 18565	− 13145	− 10945	− 11645	Sum = A	
+0·2	−0·1	−0·4	−0·4	−1·1	−1·7	−3·3	−3·4	−3·2	−2·6	−2·3	−2·7	−2·7	−2·0	−1·4	−0·7	−0·4	−0·1	$f_1 \times A$	
+2·1	−0·4	−0·9	−0·6	−1·1	−1·2	−1·6	−0·9	−0·3	−0·2	−0·6	−1·3	−1·9	−2·0	−2·0	−1·5	−1·4	−1·5	$f_2 \times A$	

Deflection in Meridian $= S - N = -73''\cdot 2$

Do. Prime Vertical $= W - E = -38\cdot 6$

KALIANA.

Height above Mean Sea Level = 814 *feet.*

Heights of Compartments in feet and Calculation of Deflection of Plumb-line. **North Side.**

Radii of Annuli		SECTORS																	
		W.			N.W.					N.					N.E.				E.
		$a' =$ 280°	290°	300°	310°	320°	330°	340°	350°	0°	10°	20°	30°	40°	50°	60°	70°	80°	90°
r'	r_1	$a_1 =$ 270°	280°	290°	300°	310°	320°	330°	340°	350°	0°	10°	20°	30°	40°	50°	60°	70°	80°
Miles 0·125	Miles 0									809	809								
0·25	0·125									810	810								
0·5	0·25									810	810								
1	0·5									810	810								
2	1									800	800								
4	2									810	810								
8	4									820	820								
16	8									830	830								
32	16	800	810	820	820	830	840	840	850	850	850	850	850	850	850	840	830	830	800
64	32	790	800	800	830	850	860	900	960	1100	1500	1500	1600	1400	1200	1600	1400	900	800
128	64	700	700	760	760	850	950	1800	4000	6000	6500	6000	6000	6000	5000	4000	4000	4000	1500
256	128	700	700	700	700	750	950	2000	5000	7000	10000	12000	12000	12000	13000	13000	11000	10000	5000
512	256	700	800	900	1700	2000	3000	8000	9000	9000	10000	13000	13000	14000	15000	15000	15000	13000	12000
1024	512	2000	4000	4000	4000	3000	4000	4000	5000	4000	4000	4000	4000	6000	8000	14000	13000	12000	6000
2048	1024	2000	800	3000	1500	300	400	500	600	800	600	1000	2000	2500	3000	3000	2000	2000	2000
Beyond		1000	0	0	0	0	0	0	0	0	0	0	0	200	200	0	*7500*	*7500*	*8000*
Sum − 12210		+2170	+2090	+4460	+3790	+2060	+4480	+11520	+18890	+22230	+26930	+31830	+32930	+36430	+39730	+44920	+33210	+28710	+13580
Correction for Curvature		−200	0	−100	−100	0	0	0	0	0	0	0	−100	−100	−100	−200	+700	+700	+800
Sum = A		+1970	+2090	+4360	+3690	+2060	+4480	+11520	+18890	+22230	+26930	+31830	+32830	+36330	+39630	+44720	+33910	+29410	+14380
$f_1 \times A$		0·0	+0·1	+0·2	+0·3	+0·2	+0·5	+1·3	+2·4	+2·9	+3·5	+4·0	+3·8	+3·9	+3·6	+3·3	+1·9	+1·0	+0·2
$f_2 \times A$		+0·3	+0·3	+0·5	+0·4	+0·2	+0·3	+0·6	+0·6	+0·2	+0·3	+1·1	+1·8	+2·7	+3·6	+4·7	+4·0	+3·7	+1·9

KALIANA.

Height above Mean Sea Level = 814 feet.

Heights of Compartments in feet and Calculation of Deflection of Plumb-line. **South Side.**

SECTORS																		Radii of Annuli	
E.				S.E.				S.				S.W.					W.		
a' =																			
100°	110°	120°	130°	140°	150°	160°	170°	180°	190°	200°	210°	220°	230°	240°	250°	260°	270°		
a_1 =																		r'	r_1
90°	100°	110°	120°	130°	140°	150°	160°	170°	180°	190°	200°	210°	220°	230°	240°	250°	260°		
								806	806									Miles 0·125	Miles 0
								805	805									0·25	0·125
								805	805									0·5	0·25
								806	806									1	0·5
								790	790									2	1
								790	790									4	2
								790	790									8	4
								760	760									16	8
770	760	760	730	740	770	760	760	770	770	740	720	740	750	760	780	790	800	32	16
790	790	790	780	770	760	750	750	740	740	740	750	760	770	770	780	780	790	64	32
1500	800	820	700	600	600	600	600	600	700	850	900	850	850	800	800	800	800	128	64
3000	1000	500	500	500	500	500	500	600	800	900	1000	1500	1500	1200	1000	700	700	256	128
5000	1200	900	400	600	900	1100	1200	1400	1500	1300	1200	1400	1000	600	800	800	400	512	256
1500	600	300	200	300	300	300	600	1000	1500	2000	500	300	300	300	0	300	1000	1024	512
2000	800	800	300	1000	4500	6500	7000	6000	4000	3000	5500	6000	6000	5000	0	0	500	2048	1024
8500	6500	1000	0	0	0	9500	8600	8500	6500	5500	5000	5500	1000	0	1000	1000	1000	Beyond	
−604	−7214	−2794	−3054	−4154	−7334	−18654	−17854	−16054	−11154	−8634	−12094	−13214	−9094	−7834	−1504	−1494	−674	Sum − 12210	
+900	+700	+100	0	0	+100	+1200	+1100	+1100	+800	+700	+700	+800	+300	+100	−100	−100	−100	Correction for Curvature	
+296	−6514	−2694	−3054	−4154	−7234	−17454	−16754	−14954	−10354	−7934	−11394	−12414	−8794	−7734	−1604	−1594	−774	Sum = A	
0·0	−0·2	−0·1	−0·2	−0·4	−0·8	−2·0	−2·1	−1·9	−1·3	−1·0	−1·3	−1·3	−0·8	−0·6	−0·1	−0·1	0·0	$f_1 \times A$	
0·0	−0·8	−0·3	−0·3	−0·4	−0·5	−1·0	−0·6	−0·2	−0·1	−0·3	−0·6	−0·9	−0·8	−0·8	−0·2	−0·2	−0·1	$f_2 \times A$	

Deflection in Meridian = S − N = − 47″·3
Do. Prime Vertical = W − E = − 20 ·3

KALIANPUR.

Height above Mean Sea Level = 1765 feet.

Heights of Compartments in feet and Calculation of Deflection of Plumb-line. North Side.

Radii of Annuli		SECTORS																	
		W.			N.W.					N.				N.E.					E.
		$a' =$ 280°	290°	300°	310°	320°	330°	340°	350°	0°	10°	20°	30°	40°	50°	60°	70°	80°	90°
r'	r_1	$a_1 =$ 270°	280°	290°	300°	310°	320°	330°	340°	350°	0°	10°	20°	30°	40°	50°	60°	70°	80°
Miles 0·125	Miles 0									1760	1760								
0·25	0·125	1750	1750	1755	1760	1760	1760	1760	1760	1760	1755	1755	1750	1750	1750	1750	1750	1750	1750
0·5	0·25	1720	1730	1735	1740	1750	1750	1750	1750	1750	1735	1730	1730	1725	1720	1720	1720	1720	1720
1	0·5	1710	1720	1730	1740	1750	1750	1750	1750	1750	1700	1700	1695	1690	1690	1700	1705	1710	1720
2	1	1700	1710	1715	1720	1730	1730	1735	1740	1740	1630	1620	1615	1610	1600	1620	1640	1660	1680
4	2	1700	1700	1700	1700	1700	1710	1715	1720	1730	1630	1620	1605	1590	1580	1580	1580	1580	1580
8	4	1680	1685	1690	1695	1700	1700	1700	1700	1700	1680	1625	1570	1515	1460	1460	1455	1450	1450
16	8	1700	1700	1700	1700	1700	1700	1700	1700	1700	1720	1675	1630	1585	1540	1520	1495	1470	1450
32	16	1700	1700	1705	1710	1710	1710	1705	1700	1700	1550	1525	1500	1475	1450	1440	1425	1410	1400
64	32	1400	1400	1405	1410	1410	1460	1505	1550	1600	1580	1560	1540	1520	1500	1475	1450	1425	1400
128	64	1200	1100	1000	1000	900	1000	1000	1200	1200	1100	900	900	900	900	900	1000	1200	1500
256	128	1500	1500	1400	1200	1100	1000	1100	1100	800	600	500	500	500	500	500	600	1100	1200
512	256	500	600	700	900	900	900	900	900	1600	4800	6000	8000	8000	6000	3000	500	300	1200
1024	512	800	2000	2900	3800	5000	5000	6000	7000	9800	11000	12000	14000	15000	15000	15000	9000	500	500
2048	1024	2000	1000	0	3000	1000	700	1500	2000	2000	2000	3000	4000	6000	5000	7000	7000	3000	2000
Beyond		1000	400	0	0	0	0	0	0	0	0	0	0	500	0	0	6000	7000	6000
Sum − 26475		−6015	−4780	−5340	−1400	−2365	−2605	−655	+1095	+4355	+8005	+10735	+15560	+18885	+15215	+14190	−155	−13200	−11925
Correction for Curvature		−200	−100	0	−100	−100	0	−100	−100	−100	−100	−100	−200	−300	−200	−300	+400	+700	+600
Sum = A		−6215	−4880	−5340	−1500	−2465	−2605	−755	+995	+4255	+7905	+10635	+15360	+18585	+15015	+13890	+245	−12500	−11325
$f_1 \times A$		−0·1	−0·2	−0·3	−0·1	−0·2	−0·3	−0·1	+0·1	+0·5	+1·0	+1·3	+1·8	+2·0	+1·4	+1·0	0·0	−0·4	−0·1
$f_2 \times A$		−0·8	−0·6	−0·6	−0·2	−0·2	−0·2	0·0	0·0	0·0	+0·1	+0·4	+0·8	+1·4	+1·4	+1·5	0·0	−1·6	−1·5

KALIANPUR.

Height above Mean Sea Level = 1765 feet.

Heights of Compartments in feet and Calculation of Deflection of Plumb-line. **South Side.**

SECTORS																		Radii of Annuli	
E.				S.E.				S.				S.W.					W.		
a′ =																			
100°	110°	120°	130°	140°	150°	160°	170°	180°	190°	200°	210°	220°	230°	240°	250°	260°	270°		
a_1 =																		r′	r_1
90°	100°	110°	120°	130°	140°	150°	160°	170°	180°	190°	200°	210°	220°	230°	240°	250°	260°	Miles	Miles
								1760	1760									0·125	0
1760	1760	1755	1750	1750	1750	1750	1750	1750	1750	1750	1750	1745	1745	1745	1740	1740	1740	0·25	0·125
1750	1750	1745	1740	1740	1740	1740	1740	1740	1740	1740	1740	1740	1740	1740	1735	1730	1730	0·5	0·25
1750	1740	1735	1730	1720	1720	1720	1720	1720	1720	1715	1710	1705	1700	1705	1710	1715	1720	1	0·5
1680	1680	1680	1680	1680	1685	1690	1695	1700	1700	1709	1700	1700	1700	1700	1695	1690	1690	2	1
1570	1570	1575	1580	1580	1600	1615	1630	1650	1690	1690	1695	1700	1700	1695	1690	1685	1680	4	2
1450	1460	1475	1490	1500	1560	1625	1690	1750	1650	1650	1650	1650	1650	1660	1665	1670	1680	8	4
1400	1410	1425	1440	1450	1490	1525	1560	1600	1650	1650	1650	1650	1650	1660	1675	1690	1700	16	8
1350	1350	1350	1350	1350	1360	1365	1370	1380	1530	1560	1590	1620	1650	1625	1600	1575	1550	32	16
1400	1500	1600	1700	1800	1760	1725	1690	1650	1500	1500	1500	1500	1500	1460	1425	1390	1350	64	32
1500	1600	1700	1700	1500	1500	1500	1800	1300	1300	1200	1600	1700	1700	1600	1500	1500	1400	128	64
1800	1800	1800	1600	1500	1400	1400	1700	1800	1800	1600	1200	1300	1300	1500	1500	1300	1200	256	128
2000	1500	1300	1200	1400	1400	1000	1000	1500	1500	1600	1900	1700	1300	300	200	500	200	512	256
100	*900*	*1500*	*2700*	*3000*	*3500*	*3500*	*1000*	1000	1500	*100*	*5000*	*4500*	*4500*	*5000*	*5000*	*5000*	*3500*	1024	512
1000	1000	*100*	*1700*	*3000*	*6500*	*8000*	*8000*	*8000*	*5000*	*7000*	*7500*	*8500*	*8500*	*8500*	*5500*	200	800	2048	1024
6000	*2000*	*900*	*0*	*0*	*6000*	*9000*	*8500*	*8500*	*8000*	*8000*	*7000*	*6000*	*7000*	*0*	1000	1000	1000	Beyond	
−11965	−9255	−9835	−11915	−13505	−23510	−28320	−24630	−22435	−18445	−22220	−26290	−25765	−27140	−21585	−17840	−12090	−10535	Sum − 26475	
+600	+200	+100	+100	+100	+800	+1200	+1100	+1100	+1000	+1100	+1000	+900	+1000	+200	+100	−100	−100	Correction for Curvature	
−11365	−9055	−9735	−11815	−13405	−22710	−27120	−23530	−21335	−17445	−21120	−25290	−24865	−26140	−21385	−17740	−12190	−10635	Sum = A	
−0·1	−0·3	−0·5	−0·9	−1·2	−2·4	−3·2	−2·9	−2·8	−2·3	−2·6	−3·0	−2·6	−2·4	−1·6	−1·0	−0·4	−0·1	$f_1 \times A$	
−1·5	−1·1	−1·1	−1·3	−1·2	−1·7	−1·5	−0·8	−0·2	−0·2	−0·7	−1·4	−1·8	−2·4	−2·3	−2·1	−1·5	−1·4	$f_2 \times A$	

Deflection in Meridian. $= S - N = -37''\cdot6$

Do. Prime Vertical $= W - E = -8\cdot5$

DAMARGIDA.

Height above Mean Sea Level = 1937 *feet.*

Heights of Compartments in feet and Calculation of Deflection of Plumb-line. **North Side.**

Radii of Annuli		SECTORS																	
		W.			N.W.					N.				N.E.					E.
		α' = 280°	290°	300°	310°	320°	330°	340°	350°	0°	10°	20°	30°	40°	50°	60°	70°	80°	90°
r'	r_1	α_1 = 270°	280°	290°	300°	310°	320°	330°	340°	350°	0°	10°	20°	30°	40°	50°	60°	70°	80°
Miles 0·125	Miles 0									1910	1910								
0·25	0·125									1910	1910								
0·5	0·25									1920	1920								
1	0·5									1920	1920								
2	1									1910	1910								
4	2									1930	1930								
8	4									1910	1910								
16	8									1850	1850								
32	16	1550	1500	1600	1800	2200	2100	1900	1800	1600	1550	1500	1500	1400	1400	1400	1600	1800	1900
64	32	1600	1700	1800	18c0	1700	1600	1500	1400	1250	1250	1100	1200	1250	1300	1400	1450	1700	2000
128	64	2000	2000	1800	1600	1450	1400	1400	14c0	1300	1250	1250	1200	110c	1200	1050	1000	1100	1400
256	128	2000	1900	2000	1800	1700	1500	1200	1500	1700	1800	1200	1000	800	900	800	1200	1200	1200
512	256	*100*	100	100	600	1000	1200	1200	1200	1600	1300	1500	1500	1700	1200	1100	1300	1500	1200
1024	512	*6000*	*5000*	*2000*	200	400	1000	1000	1600	2500	6000	7000	9000	7000	3000	800	500	*300*	*3000*
2048	1024	400	500	200	1500	3500	3500	3000	4000	4000	4500	7600	9500	10000	11000	7000	3000	1500	1000
Beyond		1000	300	0	0	0	100	200	200	300	300	700	1500	2000	2000	0	*6000*	*6000*	*7000*
Sum − 29055		−13255	−12705	−10205	−6405	−3755	−3305	−4305	−2605	−1455	+2245	+6145	+10695	+9545	+6295	−2155	−11655	−13205	−17005
Correction for Curvature		−100	0	0	0	−100	−100	−100	−100	−100	−200	−300	−500	−500	−500	−200	+600	+600	+800
Sum = A		−13355	−12705	−10205	−6405	−3855	−3405	−4405	−2705	−1555	+2045	+5845	+10195	+9045	+5795	−2355	−11055	−12605	−16205
$f_1 \times A$		−0·1	−0·4	−0·6	−0·5	−0·4	−0·4	−0·5	−0·3	−0·2	+0·3	+0·7	+1·2	+1·0	+0·5	−0·2	−0·6	−0·4	−0·2
$f_2 \times A$		−1·7	−1·6	−1·2	−0·7	−0·4	−0·3	−0·2	−0·1	0·0	0·0	+0·2	+0·6	+0·7	+0·5	−0·2	−1·3	−1·6	−2·1

DAMARGIDA.

Height above Mean Sea Level = 1937 feet.

Heights of Compartments in feet and Calculation of Deflection of Plumb-line. **South Side.**

SECTORS																		Radii of Annuli	
E.				S.E.				S.				S.W.					W.		
$\alpha' =$ 100°	110°	120°	130°	140°	150°	160°	170°	180°	190°	200°	210°	220°	230°	240°	250°	260°	270°		
$\alpha_1 =$ 90°	100°	110°	120°	130°	140°	150°	160°	170°	180°	190°	200°	210°	220°	230°	240°	250°	260°	r'	r_1
																		Miles	Miles
								1910	1910									0·125	0
								1910	1910									0·25	0·125
								1910	1910									0·5	0·25
								1910	1910									1	0·5
								1910	1910									2	1
								1910	1910									4	2
								1930	1930									8	4
								1950	1950									16	8
1800	1800	1900	1900	1900	1900	1900	1900	1900	2000	2000	2100	1950	1900	2000	2000	2000	2100	32	16
2100	2200	2100	2100	2000	2000	1900	1600	1500	1500	1500	1600	1600	1700	1800	1800	1800	1700	64	32
1500	1700	2100	2100	1800	1800	1700	1500	1300	1200	1200	1300	1350	1400	1400	1400	1700	1900	128	64
1000	1000	600	600	700	1200	900	1700	1500	1500	1500	1500	1500	1900	2000	2000	2000	2200	256	128
500	2000	3000	4000	4500	4000	300	900	2800	2800	2600	800	200	300	1600	1000	1000	1000	512	256
4500	5000	6000	6000	7000	6500	6500	0	5600	2800	3500	5000	6000	6600	6000	6500	6500	6500	1024	512
300	200	600	1200	7000	7500	8500	8500	8500	8500	5500	7600	8000	8500	8500	8500	6500	3500	2048	1024
6500	2000	0	0	5000	8500	9500	8500	8600	8000	7500	7500	6000	2500	2000	8500	2000	2000	Beyond	
−20425	−18125	−18525	−20125	−32725	−35225	−33425	−25025	−29325	−25925	−23325	−28425	−29025	−26025	−26525	−32925	−20125	−16725	Sum − 29055	
+700	+200	0	+100	+800	+1200	+1300	+1200	+1200	+1100	+1000	+1100	+900	+500	+500	+1200	0	−100	Correction for Curvature	
−19725	−17925	−18525	−20025	−31925	−34025	−32125	−23825	−28125	−24825	−22325	−27325	−28125	−25525	−26025	−31725	−20125	−16825	Sum = A	
−0·2	−0·6	−1·0	−1·5	−2·9	−3·6	−3·8	−3·0	−3·6	−3·2	−2·8	−3·2	−3·0	−2·3	−1·9	−1·7	−0·7	−0·2	$f_1 \times A$	
−2·5	−2·2	−2·2	−2·1	−2·9	−2·5	−1·8	−0·8	−0·3	−0·3	−0·7	−1·5	−2·1	−2·3	−2·8	−3·7	−2·5	−2·2	$f_2 \times A$	

Deflection in Meridian $= S - N = -38''\cdot 1$

Do. Prime Vertical $= W - E = -3\cdot 8$

PUNNÆ.

Height above Mean Sea Level = 48 *feet.*

Heights of Compartments in feet and Calculation of Deflection of Plumb-line. **North Side.**

Radii of Annuli		SECTORS																	
		W.			N.W.					N.				N.E.					E.
		α′ = 280°	290°	300°	310°	320°	330°	340°	350°	0°	10°	20°	30°	40°	50°	60°	70°	80°	90°
r′	r_1	a_1 = 270°	280°	290°	300°	310°	320°	330°	340°	350°	0°	10°	20°	30°	40°	50°	60°	70°	80°
Miles 0·125	Miles 0									48	48								
0·25	0·125									52	52								
0·5	0·25									60	60								
1	0·5									70	70								
2	1	80	80	75	75	70	72	72	75	75	75	75	70	70	65	60	50	40	30
4	2	80	90	90	90	100	100	100	100	100	100	90	90	80	70	60	60	50	50
8	4	400	500	600	500	500	300	350	360	250	250	240	130	120	110	100	90	80	70
16	8	500	800	1100	1100	1500	1600	2000	2500	250	250	200	200	150	100	40	0	*10*	*30*
32	16	0	100	200	300	1500	1800	2000	2100	1200	600	400	300	200	150	100	40	*10*	*40*
64	32	*150*	*60*	50	600	1000	2500	4000	2000	800	400	200	200	100	20	*10*	*40*	*400*	*500*
128	64	*900*	*700*	*300*	*100*	50	900	2800	4000	1200	950	500	200	80	0	*80*	*500*	*700*	*800*
256	128	*2000*	*2500*	*3500*	*2000*	*100*	0	900	2000	2000	900	700	300	100	*40*	*80*	*0*	200	500
512	256	*6000*	*3000*	*3000*	*2000*	*2500*	*400*	1000	2000	2000	2000	1200	200	*2500*	*5500*	*7000*	*6500*	*7000*	*6000*
1024	512	*9000*	*8000*	*7500*	*7500*	*8000*	*4000*	*40*	2000	1500	1000	1000	400	*3000*	*5500*	*5500*	*6000*	*6000*	*5500*
2048	1024	*6500*	*5500*	*5500*	*3000*	*3000*	*0*	1400	1400	2000	7000	6800	7000	2000	1500	1500	500	100	*500*
Beyond		2000	1500	600	500	500	300	100	400	600	800	1000	1000	1000	1000	500	0	*8500*	*8500*
Sum − 720		− 22028	− 17228	− 17623	− 11973	− 8918	+ 2634	+ 14144	+ 18397	+ 11437	+ 13787	+ 11867	+ 9552	− 2138	− 8563	− 10848	− 12838	− 22688	− 21758
Correction for Curvature		0	0	+ 100	+ 100	+ 100	0	0	− 100	− 100	− 300	− 300	− 300	− 100	− 100	− 100	0	+ 1000	+ 1000
Sum = A		− 22028	− 17228	− 17523	− 11873	− 8818	+ 2634	+ 14144	+ 18297	+ 11337	+ 13487	+ 11567	+ 9252	− 2238	− 8663	− 10948	− 12838	− 21688	− 20758
f_1 × A		−0·2	−0·6	−1·0	−0·9	−0·8	+0·3	+1·7	+2·3	+1·5	+1·7	+1·4	+1·1	−0·2	−0·8	−0·8	−0·7	−0·7	−0·2
f_2 × A		−2·8	−2·2	−2·1	−1·3	−0·8	+0·2	+0·8	+0·6	+0·1	+0·1	+0·4	+0·5	−0·2	−0·8	−1·2	−1·5	−2·7	−2·7

Height above Mean Sea Level = 48 feet.

Heights of Compartments in feet and Calculation of Deflection of Plumb-line. **South Side.**

SECTORS																		Radii of Annuli	
E.				S.E.				S.				S.W.					W.		
a' =																			
100°	110°	120°	130°	140°	150°	160°	170°	180°	190°	200°	210°	220°	230°	240°	250°	260°	270°		
a_1 =																		r'	r_1
90°	100°	110°	120°	130°	140°	150°	160°	170°	180°	190°	200°	210°	220°	230°	240°	250°	260°	Miles	Miles
								48	48									0·125	0
								48	48									0·25	0·125
								45	45									0·5	0·25
55	50	30	0	10	10	10	10	10	10	10	10	0	25	35	55	60	65	1	0·5
25	10	0	10	10	10	20	20	20	20	20	10	10	0	10	60	70	75	2	1
10	10	20	20	30	30	30	30	30	30	30	20	10	0	60	70	80	85	4	2
20	30	40	40	40	40	40	40	40	40	20	20	10	100	150	200	400	300	8	4
40	40	50	50	50	50	50	60	60	60	60	60	40	20	10	10	200	300	16	8
50	70	80	100	100	100	130	150	150	150	140	130	120	100	80	50	30	20	32	16
450	500	550	600	650	500	280	280	250	250	250	240	240	200	200	200	200	200	64	32
3000	3000	3500	4000	4000	4500	5000	4500	4000	3000	3000	3000	3000	2500	2500	2000	1500	1000	128	64
1500	2000	2000	300	3000	6000	7000	7500	7000	5000	5000	4000	4000	4000	5000	4500	4000	3000	256	128
3000	3000	4000	4000	8000	7500	7500	8500	7500	7500	8000	6600	2500	2000	2000	4000	4000	6000	512	256
5500	7000	8000	8500	8500	8500	9000	8500	7500	7500	6000	5000	6500	6500	7500	8000	7500	8500	1024	512
600	0	1500	8500	8500	9000	7000	7500	7000	6500	6500	4000	4000	4000	6500	7500	7800	7500	2048	1024
6000	500	0	0	0	5000	8500	5500	6000	5500	5500	4800	3000	0	1000	1000	800	1000	Beyond	
−17717	−12717	−16337	−26147	−33517	−41867	−45187	−43217	−40187	−36187	−35157	−28517	−24037	−19822	−25162	−27482	−24047	−25022	Sum − 720	
+700	+100	+100	+300	+300	+800	+1200	+900	+900	+800	+800	+700	+500	+100	+300	+400	+200	+100	Correction for Curvature	
−17017	−12617	−16237	−25847	−33217	−41067	−43987	−42317	−39287	−35387	−34357	−27817	−23537	−19722	−24862	−27082	−23847	−24922	Sum = A	
−0·2	−0·4	−0·9	−1·9	−3·0	−4·4	−5·1	−5·3	−5·1	−4·6	−4·3	−3·3	−2·5	−1·8	−1·8	−1·5	−0·8	−0·3	$f_1 \times A$	
−2·2	−1·6	−1·9	−2·7	−3·0	−3·0	−2·4	−1·4	−0·4	−0·4	−1·1	−1·5	−1·7	−1·8	−2·6	−3·2	−3·0	−3·2	$f_2 \times A$	

Deflection in Meridian $= S - N = -50''\cdot3$

Do. Prime Vertical $= W - E = +0\cdot7$

Height above Mean Sea Level = 30 feet.

Heights of Compartments in feet and Calculation of Deflection of Plumb-line. North Side.

Radii of Annuli		SECTORS																	
		W.			N.W.					N.				N.E.					E.
	$\alpha' =$	280°	290°	300°	310°	320°	330°	340°	350°	0°	10°	20°	30°	40°	50°	60°	70°	80°	90°
r'	r_1 / $\alpha_1 =$	270°	280°	290°	300°	310°	320°	330°	340°	350°	0°	10°	20°	30°	40°	50°	60°	70°	80°
Miles 0·125	Miles 0									30	30								
0·25	0·125									30	30								
0·5	0·25									30	30								
1	0·5	10	0	0	0	0	0	0	10	30	30	20	10	0	0	0	0	0	10
2	1	20	20	10	10	10	10	0	0	0	0	0	20	10	0	10	10	10	10
4	2	20	20	20	10	10	10	0	0	0	0	0	20	10	10	20	20	20	20
8	4	20	20	20	20	20	10	10	0	0	20	30	0	10	10	10	10	0	0
16	8	40	40	30	30	30	20	10	10	0	20	50	150	50	20	0	10	10	10
32	16	50	50	40	30	30	20	20	10	20	20	30	150	200	170	150	130	120	100
64	32	120	120	80	50	40	40	30	10	100	700	1500	1500	1500	600	500	800	1000	1200
128	64	150	150	110	80	60	60	50	50	0	500	1200	1500	2200	2000	2000	2000	2000	2000
256	128	1800	280	140	100	100	400	600	200	0	100	800	1000	1200	1000	1100	1900	1700	1600
512	256	4500	2500	1900	300	20	200	200	200	600	1400	1300	1300	1400	1300	1800	1500	1000	1300
1024	512	5000	4500	4000	1000	1500	3000	3000	3000	1500	1600	3600	4600	3500	2000	1000	1300	900	500
2048	1024	2600	2600	500	2500	3000	1000	1500	2000	4000	5600	6000	8000	13000	13000	9000	2000	2000	0
Beyond		1600	300	500	0	0	100	200	200	300	300	700	1500	2000	2000	0	6000	7000	6500
Sum − 450		−7920	−5190	−5740	+480	+3990	+4140	+4990	+5140	+6160	+9900	+14840	+19360	+24670	+21680	+15120	+3210	+1310	−220
Correction for Curvature		−200	−100	0	−100	−100	−100	−100	−100	−100	−200	−300	−400	−600	−600	−200	+600	+700	+700
Sum = A		−8120	−5290	−5740	+380	+3890	+4040	+4890	+5040	+6060	+9700	+14540	+18960	+24070	+21080	+14920	+3810	+2010	+480
$f_1 \times A$		−0·1	−0·2	−0·3	0·0	+0·4	+0·4	+0·6	+0·6	+0·8	+1·3	+1·8	+2·2	+2·6	+1·9	+1·1	+0·2	+0·1	0·0
$f_2 \times A$		−1·0	−0·7	−0·7	0·0	+0·4	+0·3	+0·3	+0·2	+0·1	+0·1	+0·5	+1·0	+1·8	+1·9	+1·6	+0·4	+0·3	+0·1

BOMBAY.

Height above Mean Sea Level = 30 feet.

Heights of Compartments in feet and Calculation of Deflection of Plumb-line. **South Side.**

SECTORS																		Radii of Annuli	
E.				S.E.				S.					S.W.				W.		
$a' =$ 100°	110°	120°	130°	140°	150°	160°	170°	180°	190°	200°	210°	220°	230°	240°	250°	260°	270°		
$a_1 =$ 90°	100°	110°	120°	130°	140°	150°	160°	170°	180°	190°	200°	210°	220°	230°	240°	250°	260°	r'	r_1
								30	30									Miles 0·125	Miles 0
								20	20									0·25	0·125
								10	10									0·5	0·25
								10	10									1	0·5
10	10	10	10	10	20	20	20	20	20	20	20	20	20	20	20	20	20	2	1
20	20	20	20	20	20	20	20	20	20	20	20	20	20	20	20	20	20	4	2
0	0	0	10	10	20	20	20	20	20	20	30	20	20	20	20	20	20	8	4
10	10	10	0	0	0	0	0	0	10	20	20	30	40	40	40	40	40	16	8
100	80	50	40	30	20	20	20	0	20	40	50	60	60	60	60	60	60	32	16
1500	1900	2000	2200	1800	1200	800	100	10	40	50	80	90	100	110	120	120	120	64	32
2000	1900	2200	2300	2700	2800	1400	200	40	70	90	110	150	170	170	160	160	160	128	64
1900	1900	1600	2000	2000	2100	2500	300	180	1500	1500	2000	2300	2300	2400	2400	2400	2000	256	128
1400	1800	1200	1300	1600	2300	2000	100	2500	2800	4600	5600	5500	6000	7000	7000	6700	6500	512	256
1500	2800	5000	4800	3000	1000	100	3000	4000	5000	7300	8000	8000	8500	8000	7100	7100	6000	1024	512
3000	5000	6000	6500	8000	7500	8000	8000	6500	7500	6500	7500	7900	8500	7500	2000	1000	1500	2048	1024
6000	5000	0	3000	6000	8000	9000	8500	8000	8000	7500	7500	6000	1000	1000	0	2000	2000	Beyond	
− 4050	− 5670	− 4400	− 6930	− 9340	− 6570	− 10670	− 19470	− 21720	− 25430	− 28090	− 31360	− 30520	− 27160	− 26770	− 19370	− 16070	− 11870	Sum − 450	
+ 700	+ 700	+ 200	+ 500	+ 900	+ 1100	+ 1200	+ 1200	+ 1100	+ 1100	+ 1000	+ 1100	+ 900	+ 400	+ 400	+ 100	− 200	− 200	Correction for Curvature	
− 3350	− 4970	− 4200	− 6430	− 8440	− 5470	− 9470	− 18270	− 20620	− 24330	− 27090	− 30260	− 29620	− 26760	− 26370	− 19270	− 16270	− 12070	Sum = A	
0·0	−0·2	−0·2	−0·5	−0·8	−0·6	−1·1	−2·3	−2·7	−3·1	−3·4	−3·5	−3·1	−2·4	−2·0	−1·1	−0·5	−0·1	$f_1 \times A$	
−0·4	−0·6	−0·5	−0·7	−0·8	−0·4	−0·5	−0·6	−0·2	−0·3	−0·9	−1·7	−2·2	−2·4	−2·8	−2·3	−2·0	−1·6	$f_2 \times A$	

Deflection in Meridian = S − N = −41″·0

Do. Prime Vertical = W − E = −20 ·3

MANGALORE.

Height above Mean Sea Level = 174 *feet.*

Heights of Compartments in feet and Calculation of Deflection of Plumb-line. North Side.

Radii of Annuli		SECTORS																	
		W.			N.W.					N.				N.E.					E.
		a' = 280°	290°	300°	310°	320°	330°	340°	350°	0°	10°	20°	30°	40°	50°	60°	70°	80°	90°
r'	r_1	a_1 = 270°	280°	290°	300°	310°	320°	330°	340°	350°	0°	10°	20°	30°	40°	50°	60°	70°	80°
Miles 0·125	Miles 0									174	174								
0·25	0·125									175	175								
0·5	0·25									175	175								
1	0·5									175	175								
2	1									170	170								
4	2									150	150								
8	4	20	20	10	10	0	0	40	100	100	110	110	120	120	130	130	150	160	160
16	8	50	50	40	40	30	10	0	100	200	200	300	300	400	500	500	400	400	400
32	16	100	90	80	70	40	20	10	50	280	500	800	900	1300	1500	1000	900	700	500
64	32	200	150	120	90	70	60	40	20	800	2000	2800	3000	3300	3500	3300	3000	2500	2000
128	64	500	450	400	300	200	100	50	0	1800	2300	2400	2300	1900	2500	3000	3000	3000	3000
256	128	1000	1000	1500	2500	2500	300	80	1000	2000	2400	2000	1500	1500	1500	1700	2000	2700	2800
512	256	5000	5000	5000	5000	5000	800	150	700	2000	1800	1400	1400	1300	1200	700	300	2000	2500
1024	512	8000	8000	7000	7000	4000	600	100	300	1000	1200	1200	1200	1200	1000	0	4000	6000	6000
2048	1024	1000	500	1000	0	2000	3000	3000	3000	3000	5000	7000	11000	12000	6000	500	500	200	500
Beyond		2000	1500	500	0	800	0	0	0	0	0	1000	3000	3000	0	1000	7000	8000	8000
Sum − 2610		− 15635	− 14525	− 14415	− 16775	− 10805	− 655	+ 1045	+ 3465	+ 9415	+ 13745	+ 17245	+ 22955	+ 24255	+ 16065	+ 8065	− 3115	− 8105	− 9905
Correction for Curvature		− 100	− 100	− 100	0	− 100	− 100	− 100	− 100	− 100	− 100	− 300	− 600	− 600	− 200	+ 100	+ 800	+ 900	+ 900
Sum = A		− 15735	− 14625	− 14515	− 16775	− 10905	− 755	+ 945	+ 3365	+ 9315	+ 13645	+ 16945	+ 22355	+ 23655	+ 15865	+ 8165	− 2315	− 7205	− 9005
$f_1 \times A$		−0·2	−0·5	−0·8	−1·2	−1·0	−0·1	+0·1	+0·4	+1·2	+1·8	+2·1	+2·6	+2·5	+1·4	+0·6	−0·1	−0·2	−0·1
$f_2 \times A$		−2·0	−1·8	−1·7	−1·8	−1·0	−0·1	+0·1	+0·1	+0·1	+0·2	+0·6	+1·2	+1·7	+1·4	+0·9	−0·3	−0·9	−1·2

MANGALORE.

Height above Mean Sea Level = 174 feet.

Heights of Compartments in feet and Calculation of Deflection of Plumb-line. **South Side.**

SECTORS																		Radii of Annuli	
E.				S.E.				S.				S.W.					W.		
α' =																			
100°	110°	120°	130°	140°	150°	160°	170°	180°	190°	200°	210°	220°	230°	240°	250°	260°	270°		
α_1 =																		r'	r_1
90°	100°	110°	120°	130°	140°	150°	160°	170°	180°	190°	200°	210°	220°	230°	240°	250°	260°	Miles	Miles
								174	174									0·125	0
								174	174									0·25	0·125
								170	170									0·5	0·25
								174	174									1	0·5
								170	170									2	1
150	150	140	140	130	120	110	100	100	100	50	10	0	0	0	10	10	10	4	2
180	160	130	100	80	60	40	20	0	0	0	10	10	10	10	10	10	20	8	4
400	400	400	300	200	100	50	0	10	10	10	20	20	30	30	40	50	50	16	8
500	700	900	900	700	700	100	10	40	60	80	90	100	100	100	100	100	100	32	16
2000	2000	2000	2000	1200	700	100	40	80	100	200	200	300	400	300	300	200	200	64	32
3000	3000	3000	3500	4500	1600	10	100	200	400	500	600	700	800	700	600	600	500	128	64
2800	2400	3000	2600	3200	2200	0	600	2500	3500	3000	2000	2500	200	200	1500	2000	2000	256	128
3000	2500	2000	100	200	1300	1200	1500	3000	5500	1500	4000	6000	5500	6500	6000	5000	5000	512	256
7000	7500	8000	7000	4500	8000	8000	7000	7500	3000	5000	5500	8000	8000	8000	8500	8500	8000	1024	512
3000	1500	1000	8000	9000	9000	8000	9000	9000	6000	6000	4000	8000	7000	8000	8000	4000	3000	2048	1024
7000	1000	0	0	0	7000	8000	6000	6000	8000	8000	6000	8000	2000	5000	1000	2000	2000	Beyond	
− 12892	− 5612	− 3352	− 7282	− 5612	− 21742	− 26732	− 26052	− 30152	− 28392	− 26162	− 24332	− 35552	− 25962	− 30762	− 25982	− 20392	− 18802	Sum – 2610	
+ 900	+ 200	+ 100	+ 200	+ 300	+ 1100	+ 1100	+ 900	+ 900	+ 1100	+ 1100	+ 800	+ 1100	+ 400	+ 800	+ 100	− 100	− 100	Correction for Curvature	
− 11992	− 5412	− 3252	− 7082	− 5312	− 20642	− 25632	− 25152	− 29252	− 27292	− 25062	− 23532	− 34452	− 25562	− 29962	− 25882	− 20492	− 18902	Sum = A	
−0·1	−0·2	−0·2	−0·5	−0·5	−2·2	−3·0	−3·1	−3·8	−3·5	−3·1	−2·7	−3·6	−2·3	−2·2	−1·4	−0·7	−0·2	$f_1 \times A$	
−1·5	−0·7	−0·4	−0·8	−0·5	−1·5	−1·4	−0·8	−0·3	−0·3	−0·8	−1·3	−2·5	−2·3	−3·2	−3·0	−2·6	−2·4	$f_2 \times A$	

Deflection in Meridian = S − N = − 41″·8

Do. Prime Vertical = W − E = − 22 ·2

MADRAS.

Height above Mean Sea Level = 54 feet.

Heights of Compartments in feet and Calculation of Deflection of Plumb-line. **North Side.**

Radii of Annuli		SECTORS: W.			N.W.					N.				N.E.					E.
	a' =	280°	290°	300°	310°	320°	330°	340°	350°	0°	10°	20°	30°	40°	50°	60°	70°	80°	90°
r'	r_1 / a_1 =	270°	280°	290°	300°	310°	320°	330°	340°	350°	0°	10°	20°	30°	40°	50°	60°	70°	80°
Miles 0·125	Miles 0									54	54								
0·25	0·125									54	54								
0·5	0·25									54	54								
1	0·5									50	50								
2	1									45	45								
4	2									45	45								
8	4	70	70	70	70	70	70	70	70	70	60	60	50	20	0	0	20	100	100
16	8	100	100	100	100	90	80	80	80	70	60	50	30	10	20	50	100	200	200
32	16	200	150	170	170	150	120	100	80	50	20	40	200	350	450	460	480	500	500
64	32	800	700	1200	1300	800	300	100	50	0	80	200	400	600	800	1000	1200	1400	1500
128	64	1700	1500	1600	1400	1100	600	300	200	30	400	800	1200	1600	2000	2500	3000	3500	4000
256	128	2800	2600	2100	1500	1200	1800	1400	800	300	0	2000	2000	3500	5000	6500	6500	6500	6500
512	256	0	1000	2000	1600	1500	1500	1300	1100	800	1100	1500	1300	500	5500	5500	6000	6000	6000
1024	512	6000	6000	6000	1300	600	800	1400	1200	1100	1300	1000	1100	300	2000	3500	4000	3500	3500
2048	1024	8500	8000	5000	4000	2000	3000	2500	3000	5000	9000	11000	12000	8000	5000	4000	3000	2000	1000
Beyond		800	1000	500	0	1000	500	0	0	0	400	1000	3000	3000	500	1000	5000	8000	6000
Sum − 810		−8592	−7442	−3822	+278	+7948	+8208	+6688	+6018	+6798	+10898	+11008	+13118	+4218	−11832	−17072	−23862	−28262	−27862
Correction for Curvature		+200	+100	+100	+100	−200	−100	−100	−100	−100	−300	−400	−600	−500	−100	0	+500	+800	+700
Sum = A		−8392	−7342	−3722	+378	+7748	+8108	+6588	+5918	+6698	+10598	+10608	+12518	+3718	−11932	−17072	−23362	−27462	−27162
$f_1 \times A$		−0·1	−0·2	−0·2	0	+0·7	+0·9	+0·8	+0·7	+0·9	+1·4	+1·3	+1·5	+0·4	−1·1	−1·3	−1·3	−0·9	−0·3
$f_2 \times A$		−1·1	−0·9	−0·4	0	+0·7	+0·6	+0·4	+0·2	+0·1	+0·1	+0·3	+0·7	+0·3	−1·1	−1·8	−2·7	−3·4	−3·5

MADRAS.

Height above Mean Sea Level = 54 feet.

Heights of Compartments in feet and Calculation of Deflection of Plumb-line. **South Side.**

SECTORS																		Radii of Annuli	
E.				S.E.				S.				S.W.					W.		
a′ =																			
100°	110°	120°	130°	140°	150°	160°	170°	180°	190°	200°	210°	220°	230°	240°	250°	260°	270°		
a_1 =																		r′	r_1
90°	100°	110°	120°	130°	140°	150°	160°	170°	180°	190°	200°	210°	220°	230°	240°	250°	260°	Miles	Miles
								54	54									0·125	0
								54	54									0·25	0·125
								54	54									0·5	0·25
								50	50									1	0·5
								40	40									2	1
								40	40									4	2
100	100	100	90	50	20	10	0	20	30	30	40	40	50	50	50	60	60	8	4
200	200	100	80	50	20	10	0	20	30	40	50	60	70	80	90	100	100	16	8
500	500	450	400	400	300	200	100	40	20	100	200	300	400	400	400	400	300	32	16
1500	1400	1200	1100	1000	900	700	500	300	150	50	100	200	300	400	500	500	700	64	32
4000	3500	3000	2500	2000	1600	1300	1000	700	400	100	100	300	500	1000	1600	1700	1600	128	64
6500	6500	7000	7000	7000	7000	6000	2000	50	10	0	100	500	1000	1400	2500	2800	2800	256	128
6000	7000	7000	7000	7000	8000	8000	1000	3000	100	3000	1500	600	800	2000	2000	1200	400	512	256
4000	5000	5000	6000	8000	8000	8500	8500	8000	8000	8000	6000	3000	4000	6000	7000	7000	7000	1024	512
0	100	500	0	8000	9000	9000	6000	8000	8000	7000	5000	6000	8000	8000	8000	9000	8000	2048	1024
3000	0	0	0	0	2000	8000	8000	7000	6000	8000	8000	7000	6000	5000	4000	1000	1500	Beyond	
−26372	−24872	−24922	−24742	−34072	−37412	−42292	−27672	−21622	−23152	−26452	−20482	−15772	−17052	−18242	−16432	−11212	−8912	Sum − 810	
+400	0	0	0	+300	+500	+1200	+1100	+1000	+900	+1100	+1000	+900	+900	+800	+700	+200	+100	Correction for Curvature	
−25972	−24872	−24922	−24742	−33772	−36912	−41092	−26572	−20622	−22252	−25352	−19482	−14872	−16152	−17442	−15732	−11012	−8812	Sum = A	
−0·3	−0·8	−1·4	−1·8	−3·1	−3·9	−4·8	−3·3	−2·7	−2·9	−3·2	−2·3	−1·6	−1·5	−1·3	−0·9	−0·4	−0·1	$f_1 \times A$	
−3·3	−3·1	−2·9	−2·6	−3·1	−2·7	−2·3	−0·9	−0·2	−0·2	−0·8	−1·1	−1·1	−1·5	−1·8	−1·8	−1·4	−1·1	$f_2 \times A$	

Deflection in Meridian = S − N = −39″·5
Do. Prime Vertical = W − E = +21 ·0

WALTAIR.

Height above Mean Sea Level = 200 feet.

Heights of Compartments in feet and Calculation of Deflection of Plumb-line. **North Side.**

Radii of Annuli		SECTORS																	
		W.			N.W.					N.				N.E.					E.
	α' =	280°	290°	300°	310°	320°	330°	340°	350°	0°	10°	20°	30°	40°	50°	60°	70°	80°	90°
r'	r_1	α_1 = 270°	280°	290°	300°	310°	320°	330°	340°	350°	0°	10°	20°	30°	40°	50°	60°	70°	80°
Miles 0·125	Miles 0									200	200								
0·25	0·125									200	200								
0·5	0·25									200	200								
1	0·5									200	200								
2	1	200	200	200	300	300	300	400	400	400	400	300	200	100	0	10	20	30	40
4	2	200	200	300	300	400	400	500	500	500	400	200	100	50	10	20	30	50	60
8	4	200	200	300	300	400	400	500	500	500	400	200	100	50	20	40	60	80	100
16	8	600	600	700	800	800	800	700	600	500	500	300	100	0	40	80	100	150	200
32	16	1200	1300	1400	1500	1500	1500	1400	1300	1200	1000	600	200	40	20	80	120	200	300
64	32	2000	2200	2500	2500	2500	3000	3000	2500	1500	1300	1000	800	300	30	80	200	350	500
128	64	1500	1300	1400	1500	1600	1400	1600	2000	2000	1700	1300	1200	1800	200	200	400	600	800
256	128	1100	1000	1000	1000	1800	1500	1500	1200	1000	800	1500	1500	1000	0	1200	3300	4500	4500
512	256	1700	1300	1200	1200	1200	1500	1600	1600	1500	1100	1400	1000	600	50	100	1000	3500	4000
1024	512	0	0	400	1000	1200	1100	1000	4000	7000	8000	9000	8000	3000	2000	2000	1000	1000	0
2048	1024	6800	4800	1000	2500	2800	2000	4000	5000	7000	8000	6000	9000	8000	5000	3000	3000	1000	0
Beyond		1000	500	400	1000	0	0	0	0	0	0	500	500	0	0	0	6000	9000	8500
Sum − 3000		+500	+1600	+6400	+11500	+12100	+11500	+13800	+17200	+20700	+21200	+19900	+20300	+12540	+4790	+790	−9630	−18860	−21400
Correction for Curvature		+100	+100	0	−200	−100	−100	−100	−200	−200	−300	−300	−300	−200	−100	−100	+600	+1000	+900
Sum = A		+600	+1700	+6400	+11300	+12000	+11400	+13700	+17000	+20500	+20900	+19600	+20000	+12340	+4690	+690	−9030	−17860	−20500
$f_1 \times A$		0·0	+0·1	+0·4	+0·8	+1·1	+1·2	+1·6	+2·1	+2·6	+2·7	+2·5	+2·3	+1·3	+0·4	+0·1	−0·5	−0·6	−0·2
$f_2 \times A$		+0·1	+0·2	+0·7	+1·2	+1·1	+0·8	+0·8	+0·6	+0·2	+0·2	+0·6	+1·1	+0·9	+0·4	+0·1	−1·1	−2·2	−2·6

WALTAIR.

Height above Mean Sea Level = 200 feet.

Heights of Compartments in feet and Calculation of Deflection of Plumb-line. **South Side.**

SECTORS																		Radii of Annuli	
E.				S.E.				S.				S.W.					W.		
a′ =																			
100°	110°	120°	130°	140°	150°	160°	170°	180°	190°	200°	210°	220°	230°	240°	250°	260°	270°		
a₁ =																		r′	r₁
90°	100°	110°	120°	130°	140°	150°	160°	170°	180°	190°	200°	210°	220°	230°	240°	250°	260°	Miles	Miles
								200	200									0·125	0
								200	200									0·25	0·125
								100	100									0·5	0·25
								100	100									1	0·5
50	40	40	40	40	30	30	20	20	20	10	0	50	100	200	200	200	200	2	1
70	80	70	70	70	60	60	50	50	40	30	20	20	100	200	200	200	200	4	2
100	130	140	120	100	90	80	70	60	50	40	30	20	100	200	300	400	500	8	4
200	250	300	300	350	300	200	150	100	100	60	40	0	100	200	400	500	600	16	8
300	300	400	400	500	400	400	400	300	300	200	200	100	60	100	500	800	1100	32	16
600	800	800	900	900	900	800	700	600	500	400	300	200	150	10	800	1400	1700	64	32
2000	2000	2000	3000	3000	2000	2000	2000	1000	1000	800	500	300	20	10	300	1000	2000	128	64
5000	5500	5000	5500	5500	6000	5000	5000	4000	4000	3000	2000	1000	400	20	400	600	1000	256	128
5000	5500	5500	6000	6000	6000	6500	6500	6500	6500	7000	3000	500	1800	1500	1700	1300	1500	512	256
1000	1500	3000	4000	5000	6500	7000	7000	7000	5000	0	1000	0	700	800	1000	1200	1700	1024	512
0	1000	0	0	0	3000	9000	7000	6000	8000	7000	5000	6000	7000	8000	8000	8000	7000	2048	1024
5000	1000	0	0	0	0	8000	8000	5000	5000	8000	7000	7000	6000	6000	8000	2000	1000	Beyond	
−21920	−20700	−19850	−22930	−24060	−27880	−41670	−39490	−33230	−33110	−29140	−21690	−17690	−14730	−14980	−14800	−7400	−1500	Sum − 3000	
+600	+100	0	0	0	+100	+1200	+1100	+700	+800	+1100	+900	+900	+800	+900	+1100	+400	+100	Correction for Curvature	
−21320	−20600	−19850	−22930	−24060	−27780	−40470	−38390	−32530	−32310	−28040	−20790	−16790	−13930	−14080	−13700	−7000	−1400	Sum = A	
−0·2	−0·7	−1·1	−1·7	−2·2	−2·9	−4·7	−4·8	−4·2	−4·2	−3·5	−2·4	−1·8	−1·3	−1·0	−0·8	−0·2	0·0	$f_1 \times A$	
−2·8	−2·6	−2·3	−2·4	−2·2	−2·1	−2·2	−1·3	−0·4	−0·4	−0·9	−1·1	−1·2	−1·3	−1·5	−1·6	−0·9	−0·2	$f_2 \times A$	

Deflection in Meridian $= S - N = -55''\cdot6$

Do. Prime Vertical $= W - E = +17\cdot5$

CALCUTTA.

Height above Mean Sea Level = 30 feet.

Heights of Compartments in feet and Calculation of Deflection of Plumb-line. **North Side.**

Radii of Annuli		W.			N.W.					N.				N.E.					E.
		SECTORS																	
		α' =																	
		280°	290°	300°	310°	320°	330°	340°	350°	0°	10°	20°	30°	40°	50°	60°	70°	80°	90°
r'	r_1	α_1 =																	
		270°	280°	290°	300°	310°	320°	330°	340°	350°	0°	10°	20°	30°	40°	50°	60°	70°	80°
Miles	Miles																		
0·125	0									30	30								
0·25	0·125									30	30								
0·5	0·25									30	30								
1	0·5									30	30								
2	1									30	30								
4	2									30	30								
8	4									30	30								
16	8									40	40								
32	16									40	40								
64	32									50	50								
128	64	200	200	150	120	120	120	110	90	70	70	60	60	60	50	50	50	50	50
256	128	1300	1500	1500	1100	800	360	150	100	100	100	100	250	700	200	100	300	500	250
512	256	1900	1500	800	300	350	2000	5000	7000	10000	10000	9000	7000	5000	2000	2000	2000	2000	2000
1024	512	1200	1200	900	600	4000	12000	14000	14000	14000	13000	12000	11000	9000	5000	4000	3000	2000	2000
2048	1024	*2000*	*0*	2000	3000	2000	5000	3000	2000	3000	3000	5000	4000	4000	2000	1000	800	500	*1000*
Beyond		1500	1000	1000	500	0	0	0	0	0	0	300	300	0	0	*1000*	*9000*	*9000*	*9000*
Sum − 450		+ 3960	+ 5260	+ 6210	+ 5480	+ 7130	+ 19340	+ 22120	+ 23050	+ 27030	+ 26030	+ 26320	+ 22470	+ 18620	+ 9110	+ 6010	− 2990	− 4090	− 5840
Correction for Curvature		− 100	− 100	− 200	− 100	− 100	− 200	− 200	− 100	− 200	− 200	− 200	− 200	− 200	− 100	+ 100	+ 1000	+ 1000	+ 1000
Sum = A		+ 3860	+ 5160	+ 6010	+ 5380	+ 7030	+ 19140	+ 21920	+ 22950	+ 26830	+ 25830	+ 26120	+ 22270	+ 18420	+ 9010	+ 6110	− 1990	− 3090	− 4840
$f_1 \times A$		0·0	+0·2	+0·3	+0·4	+0·6	+2·0	+2·6	+2·9	+3·5	+3·3	+3·3	+2·6	+2·0	+0·8	+0·5	−0·1	−0·1	−0·1
$f_2 \times A$		+0·5	+0·6	+0·7	+0·6	+0·6	+1·4	+1·2	+0·8	+0·3	+0·3	+0·9	+1·2	+1·4	+0·8	+0·6	−0·2	−0·4	−0·6

CALCUTTA.

Height above Mean Sea Level = 30 feet.

Heights of Compartments in feet and Calculation of Deflection of Plumb-line. **South Side.**

SECTORS																		Radii of Annuli	
E.				S.E.				S.					S.W.				W.		
a′ = 100°	110°	120°	130°	140°	150°	160°	170°	180°	190°	200°	210°	220°	230°	240°	250°	260°	270°		
a_1 = 90°	100°	110°	120°	130°	140°	150°	160°	170°	180°	190°	200°	210°	220°	230°	240°	250°	260°	r′	r_1
																		Miles	Miles
								30	30									0·125	0
								30	30									0·25	0·125
								30	30									0·5	0·25
								30	30									1	0·5
								30	30									2	1
								30	30									4	2
								30	30									8	4
								30	30									16	8
								30	30									32	16
								20	20									64	32
40	20	10	0	30	150	100	30	30	30	30	30	0	80	150	200	150	150	128	64
10	10	50	100	200	400	600	800	900	800	600	400	30	100	800	1600	1200	1000	256	128
1000	900	300	100	1000	2000	3500	4000	4000	4000	4000	2500	1200	300	2000	1200	1000	1000	512	256
2000	1500	900	200	50	1500	2200	5000	6000	6500	6500	5000	3000	1000	1500	1500	1400	1500	1024	512
3000	3000	200	0	100	0	500	7500	7500	7500	8500	4000	3000	5500	6000	6500	6500	6000	2048	1024
7000	7000	0	0	0	0	3000	9500	7500	7500	8500	7000	5000	4000	8000	0	0	0	Beyond	
−7160	−7780	+770	−190	−1570	−4240	−10090	−27020	−26120	−26520	−28320	−19120	−12420	−8210	−9740	−2190	−2940	−2540	Sum − 450	
+800	+800	0	0	0	0	+400	+1300	+1100	+1100	+1200	+900	+600	+600	+1000	+200	+200	+100	Correction for Curvature	
−6360	−6980	+770	−190	−1570	−4240	−9690	−25720	−25020	−25420	−27120	−18220	−11820	−7610	−8740	−1990	−2740	−2440	Sum = A	
−0·1	−0·2	0·0	0·0	−0·1	−0·4	−1·1	−3·2	−3·2	−3·3	−3·4	−2·1	−1·3	−0·7	−0·6	−0·1	−0·1	0·0	$f_1 \times A$	
−0·8	−0·9	+0·1	0·0	−0·1	−0·3	−0·5	−0·8	−0·3	−0·3	−0·9	−1·0	−0·9	−0·7	−0·9	−0·2	−0·3	−0·3	$f_2 \times A$	

Deflection in Meridian $= S - N = -44''\cdot 6$

Do. Prime Vertical $= W - E = +0\cdot 8$

BEYOND-QUETTA.

Height above Mean Sea Level = 4718 *feet.*

Heights of Compartments in feet and Calculation of Deflection of Plumb-line. **North Side.**

Radii of Annuli		SECTORS W.			N.W.					N.				N.E.					E.
		$a' =$ 280°	290°	300°	310°	320°	330°	340°	350°	0°	10°	20°	30°	40°	50°	60°	70°	80°	90°
r′	r_1	$a_1 =$ 270°	280°	290°	300°	310°	320°	330°	340°	350°	0°	10°	20°	30°	40°	50°	60°	70°	80°
Miles 0·125	Miles 0									4718	4718								
0·25	0·125									4720	4720								
0·5	0·25									4725	4725								
1	0·5									4725	4725								
2	1									4730	4730								
4	2									4735	4735								
8	4	4728	4750	4770	4765	4760	4760	4755	4755	4755	4755	4755	4750	4750	4750	4750	4750	4750	4750
16	8	5300	5500	5500	5500	5300	5300	5200	5100	5000	4800	4800	4800	4800	4800	4800	4850	4900	4900
32	16	5000	4500	4500	5000	5000	5000	5500	5600	5800	5600	5600	5500	5200	5100	5000	5000	5200	5300
64	32	3500	3500	3500	3500	3500	3800	4200	4400	5600	6200	6200	7000	7700	7300	7000	7000	6700	7500
128	64	3000	3000	3000	3500	3500	3800	4000	4200	4500	5000	5600	6000	6000	6900	6900	7000	6900	7000
256	128	2000	2300	2300	2500	3500	5000	5000	5500	6000	6000	7000	7500	8000	6000	5800	5600	5000	4000
512	256	2500	3000	3500	2800	4000	4000	5000	4000	3000	4000	5500	8000	3000	2500	1200	700	600	600
1024	512	3500	4800	3500	2000	2000	600	400	400	1000	1000	1500	3000	6500	4500	10000	13000	11000	9000
2048	1024	1000	500	3000	0	0	100	200	500	500	1000	600	1000	3500	5000	4000	15000	13000	8000
Beyond		0	0	0	0	0	0	0	0	0	0	0	0	500	0	*1000*	*2000*	*2500*	*2000*
Sum − 70770		−16607	−15285	−13565	−17570	−15575	−14775	−12880	−12680	−10980	−8780	−5580	+415	+2815	−285	+1315	+13765	+8415	+1915
Correction for Curvature		0	0	−100	0	0	0	0	0	0	0	0	0	−200	−200	−100	−200	−100	0
Sum = A		−16607	−15285	−13665	−17570	−15575	−14775	−12880	−12680	−10980	−8780	−5580	+415	+2615	−485	+1215	+13565	+8315	+1915
$f_1 \times A$		−0·2	−0·5	−0·8	−1·3	−1·4	−1·6	−1·5	−1·6	−1·4	−1·1	−0·7	0·0	+0·3	0·0	+0·1	+0·7	+0·3	0·0
$f_2 \times A$		−2·1	−1·9	−1·6	−1·9	−1·4	−1·1	−0·7	−0·4	−0·1	−0·1	−0·2	0·0	+0·2	0·0	+0·1	+1·6	+1·0	+0·2

BEYOND-QUETTA.

Height above Mean Sea Level = 4718 *feet.*

Heights of Compartments in feet and Calculation of Deflection of Plumb-line. **South Side.**

SECTORS																		Radii of Annuli	
E.				S.E.				S.				S.W.					W.		
a′ =																			
100°	110°	120°	130°	140°	150°	160°	170°	180°	190°	200°	210°	220°	230°	240°	250°	260°	270°		
a_1 =																		r'	r_1
90°	100°	110°	120°	130°	140°	150°	160°	170°	180°	190°	200°	210°	220°	230°	240°	250°	260°		
								4718	4718									Miles 0·125	Miles 0
								4710	4710									0·25	0·125
								4705	4705									0·5	0·25
								4705	4705									1	0·5
								4705	4705									2	1
								4705	4705									4	2
4800	4850	4900	5000	5000	4950	4800	4750	4750	4750	4750	4800	4800	4800	4850	4800	4800	4800	8	4
5100	5150	5200	5200	5200	5400	5400	5400	5200	5000	4700	5100	5150	5150	5150	5200	5100	5200	16	8
5400	6000	6000	5600	5400	5600	6500	5500	5700	5500	5000	5000	5200	5400	5400	5000	4800	4800	32	16
7000	7300	5500	6000	5000	4000	6000	6200	6000	6000	5500	5500	5000	4500	3000	3500	3500	3500	64	32
6000	5000	4000	2000	1000	1500	3000	5000	6600	6000	6000	5000	4000	3500	3500	3500	3500	3500	128	64
4000	3500	1800	500	200	300	200	900	2500	4000	4000	2500	2500	3800	3300	4000	3000	2000	256	128
500	500	600	700	800	600	400	300	500	0	0	500	800	2000	2000	3500	4000	3500	512	256
1500	800	1300	1500	1200	1200	0	*800*	*2500*	*6000*	*7500*	*6500*	*3000*	0	200	500	2000	3000	1024	512
800	100	*0*	*2000*	*3000*	800	0	*3000*	*8000*	*8500*	*9000*	*9000*	*3500*	0	1000	2000	1000	1000	2048	1024
8000	*3000*	*0*	*0*	*1000*	*9500*	*9000*	*8500*	*8000*	*7000*	*7000*	*9000*	*1000*	1000	2000	2000	1000	1000	Beyond	
− 20140	− 17040	− 17940	− 22740	− 27440	− 32390	− 29940	− 31490	− 34490	− 37490	− 40790	− 43340	− 27290	− 17090	− 16840	− 13240	− 14540	− 14940	Sum − 70770	
+ 900	+ 300	0	0	+ 200	+ 1000	+ 1000	+ 1000	+ 1100	+ 1000	+ 1000	+ 1300	+ 200	− 100	− 200	− 300	− 100	− 200	Correction for Curvature	
− 19240	− 16740	− 17940	− 22740	− 27240	− 31390	− 28940	− 30490	− 33390	− 36490	− 39790	− 42040	− 27090	− 17190	− 17040	− 13540	− 14640	− 15140	Sum = A	
−0·2	−0·6	−1·0	−1·7	−2·5	−3·3	−3·4	−3·8	−4·3	−4·7	−5·0	−4·9	−2·9	−1·6	−1·3	−0·7	−0·5	−0·2	$f_1 \times A$	
−2·5	−2·1	−2·1	−2·4	−2·5	−2·3	−1·6	−1·0	−0·4	−0·4	−1·3	−2·3	−2·0	−1·6	−1·8	−1·6	−1·8	−2·0	$f_2 \times A$	

Deflection in Meridian = S − N = −31″·9
Do. Prime Vertical = W − E = −11 ·9

The following results have been abstracted from the preceding tables :—

TABLE I.

Calculated Values of the Deflection of the Plumb-line.

Station	Deflection	
	In the Meridian	In the Prime Vertical
	"	"
Mussooree	−73·5	−41·1
Dehra Dún	−73·2	−38·6
Kaliána	−47·3	−20·3
Kaliánpur	−37·6	− 8·5
Dámargída	−38·1	− 3·8
Punnæ	−50·3	+ 0·7
Bombay	−41·0	−20·3
Mangalore	−41·8	−22·2
Madras	−39·5	+21·0
Waltair	−55·6	+17·5
Calcutta	−44·6	+ 0·8
Beyond-Quetta ...	−31·9	−11·9

A negative value denotes a northerly deflection in the meridian and an easterly deflection in the prime vertical.

The Disturbance of the Sea-level.

If large deflections of the plumb-line exist, they will necessarily be accompanied by considerable deformations of the level-surface: the heights of mountains and continents and the depths of seas employed as data in the preceding calculations are measured from the surface of the geoid, and consequently require a correction, if the true effect on the plumb-line of the irregular distributions of matter upon the surface of the ellipsoid of reference is to be deduced.

It is not possible to determine these corrections with exactitude, but a rough approximation can be obtained by means of Colonel Clarke's formulæ for a circular plateau.

If Y_0 and Y = elevations of the sea in feet at the centre and circumference respectively of the plateau, then

$$Y_0 = 2\,ch \sin\frac{a}{2} - 8\,ch \sin^2\frac{a}{4}$$

$$Y = 2\,ch\frac{a}{\pi} - 8\,ch \sin^2\frac{a}{4}$$

where c = radius of the earth, h = average height of plateau, a = radius of plateau.

Firstly, omitting all elevations above sea-level, we can assume India to be a submerged circular plateau of 18° in diameter and surrounded by oceans between 10000 and 11000 feet deep. The effect of the presence of water can be eliminated by multiplying the mean depth by $\frac{3}{8}$. In Colonel Clarke's formulæ the density of the attracting region is taken equal to half the mean density of the earth, and the introduction of the factor $\frac{10}{11}$ is therefore necessary to render the results consistent with the preceding calculations of the deflections, in which a mean surface density of 2·6 was employed.

$$Y_0 = \tfrac{20}{11} \times 3960 \times \tfrac{6500}{5280} \times \sin 4\tfrac{1}{2}^\circ - \tfrac{80}{11} \times 3960 \times \tfrac{6500}{5280} \times \sin^2 2\tfrac{1}{4}^\circ = 640 \text{ feet,}$$

$$Y = \tfrac{20}{11} \times 3960 \times \tfrac{6500}{5280} \times \frac{9^\circ}{\pi} - \tfrac{80}{11} \times 3960 \times \tfrac{6500}{5280} \times \sin^2 2\tfrac{1}{4}^\circ = 388 \text{ feet.}$$

Secondly, omitting the Himalaya Mountains, we can assume the continent of India, as bounded by the 100 fathom contour, to be an elevated circular plateau, of 18° in diameter, and 1800 feet in height.

$$Y_0 = 177 \text{ feet,}$$
$$Y = 108 \text{ feet.}$$

Thirdly, we can assume the plateau of Tibet to be a circle of 6° radius with an average height of 15000 feet.

$$Y_0 = 1015 \text{ feet,}$$
$$Y = 626 \text{ feet}$$

From these approximate figures the disturbance of the sea-level may be estimated to be as follows :—

At {Punnæ, Madras, Mangalore, Bombay, Waltair}	388 +	108	= 496 feet.
At Dámargída	640 +	177	= 817 feet.
At Kaliánpur	817 +	$\frac{626}{6}$	= 921 feet.
At Calcutta	817 +	$\frac{626}{4}$	= 974 feet.
At Kaliána	817 +	$\frac{3}{4} \times 626$	= 1287 feet.
At {Dehra Dún, Mussooree}	817 +	626	= 1443 feet.
In the centre of Tibet	817 +	1015	= 1832 feet.

If the above figures be accepted as indicating the disturbances of the sea-level, the corrections to the calculated values of the deflections in the meridian will be as follows* :—

TABLE II.

Corrections to calculated values of deflections in the meridian on account of disturbance of sea-level.

Station	Corrections to Calculated Values of Deflections	Difference between the correction for each station and that for Kaliánpur
	″	″
Mussooree	− 3·9	− 0·9
Dehra Dún	− 3·9	− 0·9
Kaliána	− 3·0	0·0
Kalíánpur	− 3·0	...
Dámargída	− 1·4	+ 1·6
Punnæ	− 1·1	+ 1·9
Bombay	− 1·5	+ 1·5
Mangalore.	− 1·0	+ 2·0
Madras	− 1·2	+ 1·8
Waltair	− 1·1	+ 1·9
Calcutta	− 1·7	+ 1·3

* The calculations of these corrections are as follows:—

$$\text{Deflection} = 12''\cdot44\,\frac{A}{\rho_0} = 12''\cdot44 \times \frac{2\cdot6}{5\cdot7} \times \frac{h}{5280} \times (\sin\alpha' - \sin\alpha_1) \times \log_e \frac{r'}{r_1}$$

$$= \cdot00107 \times h\ (\text{in feet}) \times (\sin\alpha' - \sin\alpha_1) \times \log_e \frac{r'}{r_1}$$

Mussooree and Dehra Dún

$$= \cdot00107 \Big\{ 389 \sin 90^\circ \log_e 3 + 389 \sin 45^\circ \log_e 3 + 500 \times 2 \sin 90^\circ \log_e 3$$

$$+ 800 \times 2 \sin 10^\circ \log_e 2 + 1200 \times 2 \sin 90^\circ \log_e 2 \Big\} = -3\cdot9$$

It is not proposed to apply the corrections shown in Table II to the values of the deflections given in Table I on page 80. The quantities in Table I have been calculated, on a certain hypothesis of density, from *known* heights and depths, and they represent, with a probability of error that can be estimated, the deflections due to visible masses. But the results in Table II have been derived from speculative data, and must be regarded not as true determinations of corrections, but as indications of the possible importance of such corrections, and of the magnitude

---

Kaliána

$$-\cdot 00107\left\{545 \sin 90^\circ \log_e 2 + 350 \times 2 \sin 90^\circ \log_e 3 + 650 \times 2 \sin 10^\circ \log_e 2 + 1060 \times 2 \sin 90^\circ \log_e 2\right\} = -3\cdot 0$$

Then, *e.g.*,

Kaliánpur

$$-\cdot 00107\left\{700 \sin 80^\circ \log_e 2 + 500 \sin 90^\circ \log_e 2 + 250 \sin 45^\circ \log_e 3 + 200 \times 2 \sin 90^\circ \log_e 2 + 350 \times 2 \sin 10^\circ \log_e \tfrac{4}{3} + 650 \times 2 \sin 90^\circ \log_e 3\right\} = -3\cdot 0$$

Dámargída

$$-\cdot 00107\left\{800 \sin 80^\circ \log_e \tfrac{3}{2} + 450 \times 2 \sin 50^\circ \log_e 2 + 100 \times 4 \sin 90^\circ \log_e 2 + 550 \times 2 \sin 20^\circ \log_e 3\right\} = -1\cdot 4$$

Punnæ

$$-\cdot 00107\left\{100 \times 2 \sin 80^\circ \log_e 10 + 600 \times 2 \sin 80^\circ \log_e \tfrac{4}{3} + 1100 \sin 40^\circ \log_e \tfrac{4}{3} + 250 \times 2 \sin 40^\circ \log_e 4\right\} = -1\cdot 1$$

Bombay

$$-\cdot 00107\left\{1100 \sin 60^\circ \log_e \tfrac{4}{3} + 400 (\sin 90^\circ + \sin 40^\circ) \log_e 3 - 100 \sin 20^\circ \log_e 2 + 250 \sin 90^\circ \log_e 4\right\} = -1\cdot 5$$

Mangalore

$$-\cdot 00107\left\{1100 \sin 40^\circ \log_e \tfrac{4}{3} + 300 \sin 40^\circ \log_e 4 + 250 \sin 90^\circ (\log_e 4 + \log_e 2) - 200 (\sin 90^\circ - \sin 50^\circ) \log_e 3\right\} = -1\cdot 0$$

Madras

$$-\cdot 00107\left\{1100 (\sin 10^\circ + \sin 80^\circ) \log_e \tfrac{4}{3} + 400 \times \sin 80^\circ (\log_e 4 + \log_e 2) + 250 \sin 90^\circ (\log_e 4 + \log_e 2) - 200 (\sin 90^\circ - \sin 60^\circ) \log_e 3\right\} = -1\cdot 2$$

Waltair

$$-\cdot 00107\left\{1100 (\sin 20^\circ + \sin 80^\circ) \log_e \tfrac{4}{3} + 300 \sin 70^\circ \log_e 3 + 250 \sin 90^\circ \log_e 4\right\} = -1\cdot 1$$

Calcutta

$$-\cdot 00107\left\{700 (\sin 40^\circ + \sin 20^\circ) \log_e 3 + 200 \times 2 \sin 90^\circ \log_e 2 + 200 \sin 90^\circ \log_e 2 + 400 (2 \sin 90^\circ - \sin 20^\circ - \sin 70^\circ) \log_e 3 + 200 (\sin 90^\circ - \sin 80^\circ) \log_e 2\right\} = -1\cdot 7$$

of the discrepancies, that it is reasonable to expect between calculated and observed results, if the disturbance of the sea-level is omitted from consideration.

Geological Considerations.

It is desirable to ascertain whether corrections should be applied to the calculated values of the deflections (page 80) on account of the differences in the rocks that constitute the surface of India. Rough approximations may be obtained by dividing India into four geological areas (Chart No. 8) :—

First area. The Himalaya Mountains, and the plateaux of Afghanistan and Tibet. This area is uncolored on Chart No. 8 and lies to the north of the yellow area.

Second area. The Indo-Gangetic Alluvium, forming the plains of Northern India from Longitude 67° to Longitude 92°, extending southwards to the parallel of 26° in the centre of India, to 24° on the west, and to 22° on the east. This area is colored yellow on Chart No. 8.

Third area. The Deccan Trap, which constitutes the surface of Western India from Latitude 16° to Latitude 25°, extending inland to the meridian of 79°. This area is colored red on Chart No. 8.

Fourth area. The great area of gneiss, that separates the trap from the alluvium in Gujarát and Bundelkhand and forms the whole of Eastern and Southern India. This area is left uncolored on Chart No. 8.

Mr. Griesbach's value for the mean density of Himalayan rocks is 2·65 and for Deccan Trap 2·95. The specific gravity of dry sand is 1·4, and of wet sand possibly 2·0. As water is found even in the desert at lesser depths than 500 feet, the density of the Indo-Gangetic Alluvium to a depth of 2000 feet may be taken to be 2·0.

The mean density of the surface of India may be estimated as follows :—

	Region	Density	No. of square degrees in area
(1).	Himalayas, Afghanistan, Tibet, ...	2·65	200
(2).	Indo-Gangetic Alluvium, ...	2·00	112
(3).	Deccan Trap, ...	2·95	40
(4).	The Gneiss area of Eastern and Southern India,	2·65	102

$$\text{Mean surface density of India} = \frac{302 \times 2\cdot65 + 112 \times 2\cdot00 + 40 \times 2\cdot95}{302 + 112 + 40} = 2\cdot51.$$

Though the mean surface density is 2·51, we may regard 2·65, which is the surface density of two-thirds of our whole area, as the *normal* surface density, and we may look upon the alluvium and trap as local deviations from the normal. We can calculate the effect of these local deviations, and show the corrections, that might be applied to the values of the deflections in Table I (page 80) on account of the existence of these areas of abnormal density.

TABLE III.

Corrections to the calculated values of deflections in the meridian on account of alluvium and trap*.

Station	Corrections to calculated values of deflections			Difference between the correction for each station and that for Kaliánpur
	On account of alluvium	On account of Trap	Total	
	"	"	"	"
Mussooree	−1·04	+0·22	−0·8	−2·7
Dehra Dún	−1·04	+0·22	−0·8	−2·7
Kaliána	−0·05	+0·31	+0·3	−1·6
Kaliánpur	+0·39	+1·52	+1·9	...
Dámargída	+0·18	−0·83	−0·7	−2·6
Punnæ	+0·05	−0·12	−0·1	−2·0
Bombay	+0·16	−0·20	0·0	−1·9
Mangalore	+0·11	−0·50	−0·4	−2·3
Madras	+0·12	−0·23	−0·1	−2·0
Waltair	+0·16	−0·08	+0·1	−1·8
Calcutta	+0·89	+0·02	+0·9	−1·0

* The alluvium has been assumed to be 2000 feet deep along its central line and the Himalayan and Vindhyan rocks on the north and south have been assumed to slope uniformly downwards under the alluvium to this maximum depth. In all probability the depth of the alluvium is greater than 2000 feet, and the slopes of the rocks steeper than have been assumed.

In the Geology of India (page 271) Mr. R. D. Oldham referring to Deccan Trap writes, "2000 feet of horizontal "beds are exposed on the flanks of Matheran Hill and a still greater thickness farther to the east in the hills near the "Bhor Ghát and close to the Great Indian Peninsula Railway line between Bombay and Poona, but it is impossible to say "how far the lowest strata, exposed at the base of the hills, are above the bottom of the series, as no lower beds than the "traps are seen. Owing to the numerous breaks in the section, it is difficult, without closer measurements than have "hitherto been made, to estimate the precise thickness of the rocks dipping to the westward near Bombay, but taking the "average dip at 5°, the whole thickness would be nearly 7000 feet. This is a minimum estimate as the average dip is "probably higher and the thickness consequently greater. From 1200 to 1500 feet of rock are exposed in Bombay island, "so that it is evident that the lowest beds seen on the island are higher in the series than the highest flows seen on the "Sahyádri Mountains to the eastward, although some of the higher portions of the range are 4000 feet above the sea."

The area colored red on Chart No. 8 has been assumed to be 4000 feet thick. The speculative nature of this assumption is recognised: it is not possible to determine from it the actual effect of Deccan Trap: we shall merely be able by means of calculations to form an idea as to whether the existence of the Trap is of importance or not.

The formula is

$$\text{Deflection} = 12''\cdot 44 \times \left(\frac{\delta - \delta_1}{\Delta}\right) \times \frac{h}{5280} (\sin \alpha' - \sin \alpha_1) \log_e \frac{r'}{r_1}$$

$$= 0''\cdot 002356 \times \left(\frac{\delta - \delta_1}{\Delta}\right) \times h \text{ (in feet)} \times (\sin \alpha' - \sin \alpha_1) \log_e \frac{r'}{r_1}$$

where $\delta = 2\cdot 65$, $\Delta = 5\cdot 7$, $\delta_1 = 2\cdot 00$ for alluvium and $2\cdot 95$ for trap.

It is not proposed to apply the corrections shown in Table III to the values of the deflections given in Table I on page 80. The results in Table III are based on mere speculations as to the depths of the alluvium and trap, and must be regarded as indications of the magnitude of the discrepancies, that it is reasonable to expect between calculated and observed results, if geological considerations are overlooked.

The effect of the Deccan Trap on the plumb-line at Kalianpur.

The persistence of the negative sign in the last column of Table III is significant: it means that the geological conditions tend to produce a more southerly deflection at Kaliánpur than at any other station, and if we examine the third column of Table III we see that the Deccan Trap exercises a strong influence at Kaliánpur. Chart No. 8 shows that Kaliánpur is situated at the north-east corner of the red or trappean area, and the following questions at once occur: Can this trappean area be causing the southerly and westerly deflections at Kaliánpur, which Walker and Strahan have deduced (page 4) from the observations of all India? Can this area of trap be causing a southerly deflection at all the stations of Lenox Conyngham's group? Can the Deccan Trap be the hidden cause of the "Mean of India" system of deflections (page 40) and of the excess of negative values of (O — C) in India (page 4)?

Assuming the trap to have an uniform depth of 4000 feet, we can calculate its effect on the stations of Lenox Conyngham's group*.

The results of a calculation are as follows :—

Station and distance from Kaliánpur		Deflection due to Trap	Difference from Kaliánpur
		″	″
Daiádhari ...	35 miles north	+2·37	+0·80
Súrantál ...	8 miles north	+1·69	+0·12
Kaliánpur ...	...	+1·57	...
Kámkhera ...	9 miles south	+1·49	−0·08
Ahmadpur ...	35 miles south	+1·26	−0·31

Dehra Dún and Mussooree.

Alluvium

$$-0''\cdot002356 \times \frac{2\cdot65 - 2\cdot00}{5\cdot7} \times 1000 \times \left\{2 \sin 40^\circ \left(\log_e \frac{200}{20} + \log_e 2\right)\right\} = -1''\cdot04$$

Trap

$$+0''\cdot002356 \times \frac{2\cdot95 - 2\cdot65}{5\cdot7} \times 4000 \times \sin 30^\circ \times \log_e \tfrac{5}{2} = +0''\cdot22$$

$$\text{Total} = -0''\cdot8$$

* This effect largely depends on whether the trap ends abruptly or gradually, and whether at its junction with the surrounding gneiss it underlies or overlies the latter. Our uncertainty on these points renders it impossible for us to make even an approximate estimate of the deflections at stations situated on the very border of the trap.

These results, based though they are on a speculative hypothesis of depth, suffice to show, that astronomical observations taken at a group of stations, situated in one corner of an area of trap, do not in their mean give an absolute value of latitude, more reliable or more free from local attraction than observations at the central station only. *In such a situation a group is futile: it not only fails to eliminate the effects of local attraction, but it fails to warn us of the existence of local attraction.*

Assuming a depth of 4000 feet we have shown that the trappean area may produce a southerly deflection at Kaliánpur of 1″·57: if we assume a depth of 6000 feet, the resulting southerly deflection will theoretically be 2″·36: and if the depth is taken as 1000 feet, the theoretical deflection will be 0″·39. We can perhaps test the *actual* effects of the trap in the following way:—observations for latitude have been taken at every 40, 50 or 60 miles on the meridians of 74°, 75°, 76° and 77° from north to south across the trap. If this trappean mass is exercising a paramount effect, the observations at stations on the northern edge of the area should indicate a southerly deflection in the meridian, at stations on the southern edge a northerly deflection, and at stations in the heart of the trap no deflection. From the table following page 14, we abstract the following apparent values of deflections in the meridian:—

At stations near the northern edge of the Trap:

Station	Deflection	
Daiádhari	+ 1″·01	Mean − 1″·2
Guraria	− 0 ·79	
Aramlia	− 4 ·92	

At stations near the central parallel of the Trap:

Station	Deflection	
Colába	− 10″·64	Mean − 8″·0
Valvádi	− 6 ·77	
Kanheri	− 9 ·12	
Badgaon	− 7 ·83	
Voi	− 5 ·51	

At stations near the southern edge of the Trap:

Station	Deflection	
Majala	− 1″·68	Mean − 2″·1
Mávinhúnda	− 0 ·03	
Dámargída	− 2 ·74	
Kodangal	− 3 ·92	

These quantities are differential, and are affected by a constant error equal in amount to the deflection caused by the trap at Kaliánpur: but whatever southerly deflection we assume to exist at Kaliánpur, the broad fact remains that a belt of maximum northerly deflections crosses the centre of the trappean area from west to east. Over the heart of the trap, where theory places the minimum deflections, we find the maximum values: and along the southern edge of the trap, where theory places the maximum northerly deflections, we find the values to be considerably less than those of the central parallel: though therefore the calculations have shown that the results of the Kaliánpur group may have been vitiated by the presence of the trap, yet the latitude observations taken throughout the trappean area give no confirmation to the theory.

The belt of maximum values, that crosses the heart of the trap from west to east, denotes either that the depth of the northern portion of the trap is many miles in excess of that of the southern, or that the effect of the trap, whatever it may be, is masked by more powerful influences.

The northern positive zone and the Indo-Gangetic Alluvium.

The surface of the rocky area to the south of the Indo-Gangetic alluvium is higher by 1000 feet* than that of the alluvium itself. This superior elevation tends to produce southerly deflections along the southern border of the alluvium.

Table III on page 85 shows that the defective density of the alluvium, apart from any question of height, may cause a *southerly* deflection at Kaliánpur of 0″·39 : Kaliánpur is 150 miles south of the alluvium, and the effects of the latter's inferior elevation and defective density will gradually increase as its southern edge is approached from Kaliánpur : *the alluvium will therefore tend to produce a zone of positive values of (O – C) in latitude over the area, where the northern positive zone actually exists* (Chart No. 6). Can the northern positive zone, described on page 14, be due then to the Indo-Gangetic alluvium? The following are the calculated effects on the plumb-line of this alluvium :—

	Southerly Deflections due to the alluvium			Difference from Kaliánpur
	On account of Inferior Elevation	On account of Defective Density	Total	
	″	″	″	″
At Kaliánpur	+ 1·3	+ 0·4	+ 1·7	...
At 50 miles north of Kaliánpur ...	+ 1·4	+ 0·5	+ 1·9	+ 0·2
At 100 miles north of Kaliánpur ...	+ 1·6	+ 0·6	+ 2·2	+ 0·5
At 150 miles north of Kaliánpur ...	+ 1·4	+ 1·7	+ 3·1	+ 1·4
At 200 miles north of Kaliánpur ...	+ 0·6	+ 1·6	+ 2·2	+ 0·5
			Mean ...	+ 0″·7

The mean value of (O – C) in latitude, derived from actual observations, within the northern positive zone is + 1″·04 (page 21). It is clear then that in the Indo-Gangetic alluvium we have found a possible cause of the northern positive zone. Whether it is the actual cause, we cannot decide, until we have arrived at more definite conclusions on collateral questions : so long as we hold that deflections at coast stations are *towards* the ocean, we shall not be able to accept any hypothesis denying the existence of deflections *towards* the alluvium : the Indo-Gangetic alluvium fills a deep and wide channel, that was possibly once a branch of the ocean : we cannot at one moment assume the ocean to be compensated by underlying strata of excessive density, and at another moment assume the alluvium to be uncompensated. We cannot apply contradictory theories to different data : we must classify our results, and treat them from one standpoint as parts of one whole.

Dehra Dún :
October 1901.

* South of Bengal the rocky area attains elevations 2000 feet above the alluvium.

PART V.

Comparison of calculated with observed values of Deflections in the Meridian.

Results of Observation.

The results of the Indian observed latitudes require to be corrected for the heights of stations above sea level, before a comparison with calculated values is instituted. In Part I the expression (O − C) was used to denote the difference between astronomical and geodetic values: the astronomical values were designated O, as being observed, the geodetic C as being computed by the usual geodetic formulæ through the triangulation. The investigations in Part IV have now supplied a *third* value, derived from a calculation of the effects of masses: to avoid confusion we will in future denote the *observed* or *astronomical* value by A, and the *geodetic* value by G, and substitute the form (A − G) for (O − C): the term "computed value" will not be used, and the "calculated value" will invariably mean the value deduced from a calculation of the effects of visible *uncompensated* masses.

TABLE IV.

SHOWING VALUES OF DEFLECTIONS AS DEDUCED FROM OBSERVATION.

Station	Observed Latitude	Correction to Sea-level	Seconds of Corrected Observed Latitude = A	Seconds of Geodetic Latitude = G	(A − G) on Everest Spheroid
	° ′ ″	″	″	″	″
Mussooree	30 27 4·02	−0·31	3·71	40·79	−37·08
Dehra Dún*	30 18 51·92	·10	51·82	88·97	−37·15
Kaliána	29 30 47·98	·04	47·94	54·94	− 7·00
Kaliánpur	24 7 11·57	·07	11·50	...	...
Dámargída	18 3 14·92	·06	14·86	17·59	− 2·73
Punnæ	8 9 29·92	·00	29·92	28·03	+ 1·89
Bombay	18 53 39·16	·00	39·16	49·72	−10·56
Mangalore	12 52 17·76	·00	17·76	15·00	+ 2·76
Madras Observatory† ...	13 4 8·0	·00	8·0	4·40	+ 3·6
Waltair	17 43 20·38	·01	20·37	29·55	− 9·18
Calcutta	22 32 55·58	·00	55·58	54·91	+ 0·67

The correction for height reduces the fundamental latitude of Kaliánpur (page 7) to 24° 7′ 11″·50 and all geodetic latitudes by 0″·07.

* The *new* latitude station at Dehra, *vide* note on page 58.

† See Madras Meridian Circle Observations, Volume IX, page XXI. The table following page 14 was printed, before the Madras Volume had been received: in future Mr. Michie Smith's value for the astronomical latitude of Madras will be adopted in the Trigonometrical Survey.

Results of Calculation.

The differences between the calculated values of the deflections at the several stations and that at Kalíánpur are given in Table V.

TABLE V.

Station	Calculated Deflection in the Meridian from Table I. (p. 80) = S	Calculated Deflection at Kalíánpur = K	Difference = S − K
	"		"
Mussooree ...	−73·5		−35·9
Dehra Dún ...	−73·2		−35·6
Kaliána	−47·3		− 9·7
Kalíánpur... ...	−37·6		...
Dámargída ...	−38·1	" −37·6	− 0·5
Punnæ	−50·3		−12·7
Bombay	−41·0		− 3·4
Mangalore ...	−41·8		− 4·2
Madras	−39·5		− 1·9
Waltair	−55·6		−18·0
Calcutta	−44·6		− 7·0

The results in Table V show that the calculated value of the deflection at Kalíánpur is less than at any other station, and that apparent northerly deflections and negative values of (A − G) in latitude may consequently be expected to preponderate in every part of India (page 7).

Comparison of results of Calculation and Observation.

The comparison between the results of calculation and observation is made in Table VI.

TABLE VI.

Station	Calculated Deflection in the meridian from Table V — Other Stations − Kalíánpur	(A − G) in Latitude from Table IV	Discrepancy between calculation and observation
	"	"	"
Mussooree ...	−35·9	−37·1	+ 1·2
Dehra Dún ...	−35·6	−37·2	+ 1·6
Kaliána	− 9·7	− 7·0	− 2·7
Kalíánpur... ...	...	...	...
Dámargída ...	− 0·5	− 2·7	+ 2·2
Punnæ	−12·7	+ 1·9	−14·6
Bombay	− 3·4	−10·6	+ 7·2
Mangalore ...	− 4·2	+ 2·8	− 7·0
Madras	− 1·9	+ 3·6	− 5·5
Waltair	−18·0	− 9·2	− 8·8
Calcutta	− 7·0	+ 0·7	− 7·7

When considering the discrepancies between calculation and observation we have to bear in mind that the calculated values given in the second column of Table VI depend on an *assumed* ratio of surface to mean density, and that the observed values given in the third column of Table VI are based on an *assumed* ellipsoid of reference.

Uncertainties arising from the adopted ratio of density.

In Table VII are given the calculated values of the deflections on different assumptions of the ratio of surface to mean density.

TABLE VII.

Station	Calculated Deflections in the meridian, if the ratio of surface to mean density is			Difference between the deflection at Kaliánpur and that at other stations, if the density-ratio is		
	$\frac{1}{2\cdot0}$	$\frac{1}{2\cdot2}$	$\frac{1}{2\cdot4}$	$\frac{1}{2\cdot0}$	$\frac{1}{2\cdot2}$	$\frac{1}{2\cdot4}$
	"	"	"	"	"	"
Mussooree	−80·9	−73·5	−67·4	−39·5	−35·9	−32·9
Dehra Dún	−80·5	−73·2	−67·1	−39·2	−35·6	−32·6
Kaliána	−52·0	−47·3	−43·4	−10·7	− 9·7	− 8·9
Kaliánpur	−41·4	−37·6	−34·5	...	...	...
Dámargída	−41·9	−38·1	−34·9	− 0·6	− 0·5	− 0·5
Punnæ	−55·3	−50·3	−46·1	−14·0	−12·7	−11·6
Bombay	−45·1	−41·0	−37·6	− 3·7	− 3·4	− 3·1
Mangalore	−46·0	−41·8	−38·3	− 4·6	− 4·2	− 3·8
Madras	−43·5	−39·5	−36·2	− 2·1	− 1·9	− 1·7
Waltair	−61·2	−55·6	−51·0	−19·8	−18·0	−16·5
Calcutta	−49·1	−44·6	−40·9	− 7·7	− 7·0	− 6·4

Uncertainties due to the Ellipsoid of reference.

In Table VIII are given the observed values of the deflections, as deduced from different Ellipsoids of reference.

TABLE VIII.

	Observed Deflections in the Meridian = (A − G) in latitude		
	On the Everest Spheroid	On the Clarke Spheroid	On a third Spheroid
Major axis in feet......	20, 922, 932	20, 926, 202	20, 926, 202
Ellipticity...........	$\frac{1}{300\cdot80}$	$\frac{1}{293\cdot47}$	$\frac{1}{300\cdot80}$
	"	"	"
Mussooree	−37·1	−35·8	−33·5
Dehra Dún	−37·2	−36·2	−33·6
Kaliána	− 7·0	− 6·2	− 3·9
Kalíánpur	...	...	...
Dámargída	− 2·7	− 3·3	− 6·2
Punnæ	+ 1·9	+ 1·2	− 7·2
Bombay	−10·6	−11·1	−13·6
Mangalore	+ 2·8	+ 2·1	− 3·6
Madras	+ 3·6	+ 2·9	− 2·7
Waltair	− 9·2	− 9·8	−12·8
Calcutta	+ 0·7	+ 0·4	− 0·2

It is interesting to see from this Table that the adoption of the third spheroid would convert the positive value of (A − G) that now exists at Punnæ into a large negative value and would consequently eliminate the *southern* positive zone (page 20).

Degree of Uncertainty attaching to the comparison of calculated and observed values.

Tables VII and VIII show the degree of uncertainty attaching to the figures of Table VI. The comparison between the results of calculation and observation, as made in Table VI, is repeated in Table IX on nine different hypotheses, each value of the density-ratio assumed in Table VII being successively combined with each of the three ellipsoids of Table VIII. In Table IX a *negative* discrepancy implies that the calculated value of the deflection is larger than the observed.

TABLE IX.

	Discrepancies between results (Calculated – Observed)								
	Everest Spheroid			Clarke Spheroid			Third Spheroid		
Density-Ratio	$\frac{1}{2\cdot0}$	$\frac{1}{2\cdot2}$	$\frac{1}{2\cdot4}$	$\frac{1}{2\cdot0}$	$\frac{1}{2\cdot2}$	$\frac{1}{2\cdot4}$	$\frac{1}{2\cdot0}$	$\frac{1}{2\cdot2}$	$\frac{1}{2\cdot4}$
	"	"	"	"	"	"	"	"	"
Mussooree ...	− 2·4	+ 1·2	+ 4·2	− 3·7	− 0·1	+ 2·9	− 6·0	− 2·4	+ 0·6
Dehra Dún ...	− 2·0	+ 1·6	+ 4·6	− 3·0	+ 0·6	+ 3·6	− 5·6	− 2·0	+ 1·0
Kaliána ...	− 3·7	− 2·7	− 1·9	− 4·5	− 3·5	− 2·7	− 6·8	− 5·8	− 5·0
Kaliánpur ...	...	...	...	...	...	...	...	...	...
Dámargída ...	+ 2·1	+ 2·2	+ 2·2	+ 2·7	+ 2·8	+ 2·8	+ 5·6	+ 5·7	+ 5·7
Punnæ • ...	−15·9	−14·6	−13·5	−15·2	−13·9	−12·8	− 6·8	− 5·5	− 4·4
Bombay ...	+ 6·9	+ 7·2	+ 7·5	+ 7·4	+ 7·7	+ 8·0	+ 9·9	+10·2	+10·5
Mangalore ...	− 7·4	− 7·0	− 6·6	− 6·7	− 6·3	− 5·9	− 1·0	− 0·6	− 0·2
Madras ...	− 5·7	− 5·5	− 5·3	− 5·0	− 4·8	− 4·6	+ 0·6	+ 0·8	+ 1·0
Waltair ...	−10·6	− 8·8	− 7·3	−10·0	− 8·2	− 6·7	− 7·0	− 5·2	− 3·7
Calcutta ...	− 8·4	− 7·7	− 7·1	− 8·1	− 7·4	− 6·8	− 7·5	− 6·8	− 6·2
Mean discrepancy	− 4·7	− 3·4	− 2·3	− 4·6	− 3·3	− 2·2	− 2·5	− 1·2	− 0·1

The discrepancies, exhibited in Table IX, between calculated and observed results are neither large nor persistent in sign, and neither require nor support in themselves the theory of uniform compensation. But their smallness is possibly unreal and misleading: the attraction of the Himalayas and the repulsion of the Ocean, if uncompensated, conspire to produce large but *parallel* deflections in the meridian throughout India. Comparisons cannot be made between calculated and observed values of *absolute* deflections: we have to be content with comparing the calculated *differences* of deflections from Kaliánpur with the observed *differences*; observed differences of deflections are almost always *small* in all parts of the world and calculated differences will be *rendered* small, if the active forces tend to produce *parallel* deflections. We cannot thus attach great weight to an accordance between calculated and observed values of small meridional deflections in India: Table IX shows in fact an absence of discordance rather than an existence of accordance; it shows that the theory of Himalayan compensation must depend for its support on the results of longitude and pendulum observations and not on discrepancies in latitude.

It must however be admitted that the agreement at Dehra Dún and Mussooree is noteworthy: nature rarely exhibits great deflections on the earth's surface, and that great deflections should be found in accordance with the laws of gravitation, is a remarkable and significant fact.

Analysis of the Calculated Values.

We have so far been dealing with the *total* values of the deflections, as calculated, without examining their composition: it will, however, facilitate future investigation, if we analyse the total values and ascribe the due component parts to the respective sources of attraction. The heights of the compartments have been entered in the Tables (pages 56 to 79) in different types: Himalayan and Oceanic areas are thus easily identified: the heights of Continental India and Further Asia are given in a like type, but as a great difference of elevation occurs at the boundary between Afghanistan and the Punjab, their dividing line on the Tables is easily followed. The deflections due respectively to the Himalaya Mountains, the Indian Ocean, Continental India, and Further Asia have been calculated and are given in Table X.

TABLE X.

Deflections due to the Himalayas, the Ocean, Continental India and Further Asia *at points on the sea-level vertically below* stations of observation.

Stations	Deflections due to				Total Deflections
	Himalayas	Ocean	India	Further Asia	
	"	"	"	"	"
Mussooree	−64·9	−10·3	+ 5·0	− 3·3	−73·5
Dehra Dún	−72·2	−10·3	+12·6	− 3·3	−73·2
Kaliána	−36·2	−11·0	+ 1·9	− 2·0	−47·3
Kaliánpur	−18·4	−19·4	+ 3·1	− 2·9	−37·6
Dámargída	−10·0	−26·2	+ 0·1	− 2·0	−38·1
Punnæ	− 3·4	−37·6	− 8·7	− 0·6	−50·3
Bombay	− 7·9	−29·6	− 1·3	− 2·2	−41·0
Mangalore	− 4·9	−28·6	− 6·3	− 2·0	−41·8
Madras	− 6·8	−28·0	− 3·6	− 1·1	−39·5
Waltair	−11·0	−33·0	−10·9	− 0·7	−55·6
Calcutta	−23·3	−19·9	− 0·4	− 1·0	−44·6

Chart No. 10 was plotted from the data furnished by Table X, and exhibits the respective uncompensated effects of the Himalaya Mountains, of the Indian Ocean, and of Continental India on the plumb-line at all points on the Great Arc of India from Mussooree to Punnæ.

The effect of the Indian Ocean has been analysed, and in Table XI are shown the negative effects on the plumb-line of the waterless void, and the positive effects of the great mass of sea-water.

TABLE XI.

EFFECTS OF THE OCEAN.

Station	Deflection due to vacant depths of Ocean	Deflection due to the water in the Ocean	Resultant effect of Ocean
	"	"	"
Mussooree	−17·2	+ 6·9	−10·3
Dehra Dún	−17·2	+ 6·9	−10·3
Kaliána	−18·3	+ 7·3	−11·0
Kaliánpur	−32·3	+12·9	−19·4
Dámargída	−43·7	+17·5	−26·2
Punnæ	−62·7	+25·1	−37·6
Bombay	−49·3	+19·7	−29·6
Mangalore	−47·7	+19·1	−28·6
Madras	−46·7	+18·7	−28·0
Waltair	−55·0	+22·0	−33·0
Calcutta	−33·2	+13·3	−19·9

Dehra Dún:

October 1901.

PART VI.[1]

Comparison of calculated with observed values of Deflections in the Prime Vertical.

In Table XII are shown the discrepancies between the calculated and observed values of the deflections in the prime vertical.

TABLE XII.

Station	Absolute Calculated Values from Table I	Calculated Deflection in the Prime Vertical — Other Stations — Kaliánpur	(A − G) in Longitude × cos φ	(A − G) in Azimuth × cot φ	Discrepancy between calculation and observation
	"	"	"	"	"
Mussooree	−41·1	−32·6	...	−26·0	− 6·6
Dehra Dún	−38·6	−30·1	−22·1	−23·0	− 8·0
Kaliána	−20·3	−11·8	...	− 4·4	− 7·4
Kaliánpur	− 8·5	...	...	...	...
Dámargída	− 3·8	+ 4·7	...	− 9·8	+14·5
Punnæ	+ 0·7	+ 9·2	− 1·8	...	+11·0
Bombay	−20·3	−11·8	+ 6·4	...	−18·2
Mangalore	−22·2	−13·7	+ 1·9	...	−15·6
Madras	+21·0	+29·5	− 7·0	...	+36·5
Waltair	+17·5	+26·0	− 3·1	...	+29·1
Calcutta	+ 0·8	+ 9·3	−10·1	...	+19·4

It is clear, that the discrepancies in the prime vertical, with which we have now to deal, are more serious than those of the meridian (in Part V). The observations at all stations excepting Mussooree, Dehra Dún and Kaliána now give a deflection *in the opposite direction to that which the theory of gravitation requires.*

The trans-continental Arcs of India.

The details of three arcs of longitude that cross India from east to west are given in Table XIII.

TABLE XIII.

Arc of Longitude	In Latitude	Astronomical value = A	Seconds of Geodetic value = G	Observed deflections = (A − G) On the Everest Spheroid	On the Clarke Spheroid
	°	m s	s	"	"
Amritsar–Mooltan ...	31	13 44·285	43·737	+ 8·22	+10·43
Waltair–Bombay	18	42 0·290	0·961	−10·06	− 3·84
Madras–Mangalore ...	13	21 36·157	36·775	− 9·27	− 6·15

A negative value of (A−G) denotes *outward* deflections at the terminals of an arc, a positive value *inward* deflections. Amritsar and Mooltan are in the plains of the Punjab, (Chart No. 11): the Himalayas rise 80 miles east of Amritsar, the Baluchistan plateau rises 70 miles west of Mooltan: there are thus mountain masses on the outer flanks of the two terminals of this arc: the ground that intervenes between Amritsar and Mooltan is the low-lying alluvium. The astronomical value of the arc is *greater* than the geodetic, and the plumb-lines are apparently deflected *inwards, from* the hills *towards* the alluvium.

Madras and Mangalore are on the coast (Chart No. 11) and a high plateau intervenes: the astronomical value of this arc is *smaller* than the geodetic, and the plumb-lines are apparently deflected *outwards, from* the plateau *towards* the ocean.

Waltair and Bombay are on the coast (Chart No. 8): the astronomical value of the Waltair-Bombay arc is *less* than the geodetic, and the plumb-lines are apparently deflected seawards.

It will be seen from Table XIII that no modification of the ellipsoid of reference will produce accordance between astronomical and geodetic values: if we adopt the Clarke spheroid to suit the southern arcs, we increase the value of (A − G) at Amritsar and Mooltan; if we determine a spheroid to fit the northern arc, we find ourselves confronted with enhanced discrepancies on the southern.

On the three trans-continental arcs the plumb-lines are thus deflected in the opposite direction to that, which the theory of gravitation would lead us to expect: in the following table are given the actual numerical discrepancies between the results of calculation and observation.

TABLE XIV.

A negative value denotes *outward* deflections at the terminals of an arc, a positive value *inward* deflections.

Arcs of Longitude	Difference between the Astronomical and Geodetic values of the arcs		Discrepancy
	As deduced theoretically by calculating the attractive effects of masses	As derived practically from observation	
	″	″	″
Amritsar–Mooltan ...	−20*	+ 8·22	−28
Waltair–Bombay ...	+39·6	−10·06	+49
Madras–Mangalore ...	+44·7	− 9·27	+53

On each arc the results of observation favour a *theory of entire compensation.*

Dehra Dún:

November 1901.

* Estimated.

PART VII.

It is inferred that a hidden cause in Central India is masking true Himalayan effects.

The seaward deflections at coast stations.

The longitude observations show seaward deflections in the prime vertical at Bombay, Waltair, Madras and Mangalore: the latitude observations show northerly deflections at Bombay and Waltair, and southerly at Madras and Mangalore: it is of interest to ascertain the *resultant* directions of deflections at the four stations.

TABLE XV.

Station	Coast	Deflections of the Plumb-line	
		In Latitude (p. 92.)	In Longitude (pp. 15 and 96.)
Bombay ...	West	−11″ North	+6″ West
Waltair ...	East	− 9 North	−3 East
Madras... ...	East	+ 4 South	−7 East
Mangalore ...	West	+ 3 South	+2 West

The resultant deflection at Bombay is towards the N.N.W., that at Waltair towards the N.N.E., that at Madras E.S.E. and that at Mangalore S.S.W. At each corner of the quadrilateral the plumb-line is being pulled diagonally outwards. The deflection at Madras is towards the deepest part of the Bay of Bengal, that at Mangalore towards the Laccadive coral-reefs. The deflection at Bombay is *towards the shallows of the Gulf of Cambay, and the land of Kathiawar.* The resultant deflection at Waltair is *towards the Orissa Coast and the shoals at the mouths of the Ganges.* If the seaward deflections of the plumb-line in the prime vertical are really due to the contraction and condensation of the strata under the bed of the Ocean, and to the attraction of this excess of submarine matter, a seaward deflection should also be perceptible to observation in the meridian. If the depths of the Arabian Sea and Bay of Bengal are the centres of attraction, the plumb-lines at Bombay and Waltair should be deflected respectively towards the S.W. and S.E. On the other hand, if continental India produces the great northerly deflections, that have been observed in the meridian at Bombay and Waltair, the same continental India should cause landward deflections in the prime vertical. The testimony is in fact conflicting, and one cannot but infer the existence of an additional and undiscovered source of error or attraction which is masking the true effects of ocean and continent. Diagonal outward deflections tend to throw suspicion on the adopted ellipsoid of reference.

The Conflict of Evidence.

In Part IV attention was drawn to the contradictory effects of mountains and seas (p. 47). In Parts V and VI we found ourselves confronted with further contradictions: an examination of the tables of Part V led to the conclusion, that the Himalaya Mountains and the Indian Ocean were not compensated to any considerable extent: the numerical results of the tables of Part VI justified the inference, that the Himalaya Mountains and the Indian Ocean were wholly compensated. The results of the latitude observations at Bombay, Waltair, Dehra Dún, Mussooree, Nojli, etc., seemed to denote an absence of compensation: the results of the longitude observations at Amritsar, Mooltan, Bombay, Waltair, etc., appeared to show that the compensation was everywhere entire. Whether the theory of uniform compensation be correct, or incorrect, there should be no *direct* conflicts of testimony: *the true inferences to be drawn from the contradictions in the evidence are, that there is an undiscovered cause at work affecting our numerical results, and that the theory of compensation is not in itself sufficient to account for the observed phenomena.*

In the following table the evidence is summarised.

TABLE XVI.

SUMMARY OF AVAILABLE EVIDENCE.

Observed Phenomena			Explanation of Phenomena, if mountains and seas are assumed to be	
			Not compensated	Completely compensated
From latitude observations.	(1)	Persistence of the negative sign in the values of (A — G) in latitude between the parallels of 14° and 24°.	The plumb-line is deflected to the north throughout India by the attraction of the Himalayas and the repulsion of the Ocean, their combined effect being a minimum at Kalíánpur.	The plumb-line at Kalíánpur is deflected to the south and thus all geodetic latitudes are too large.
	(2)	The belt of negative maxima that crosses India from west to east in latitude 20° (*vide* Table following page 14).	The Vindhya Mountains cause northerly attraction and (this cause being by itself insufficient) the conjunction of Himalayan attraction with oceanic repulsion tends perhaps to create maxima within this belt.	No explanation.
	(3)	The northern positive zone in latitude 25°.	The attraction of the Vindhya and Aravalli Mountains and of the Deccan Trap draws the plumb-line away from the Gangetic alluvium.	No explanation.
	(4)	The southern positive zone between the parallels of 8° and 13°.	The errors of the Clarke and Everest spheroids produce this zone.	The submarine strata attract the plumb-line seawards.
	(5)	The large deflections in Sub-Himalayan Regions.	The Himalayas, if uncompensated, would produce these deflections.	No explanation.
From longitude observations.	(6)	The seaward deflections in longitude at all coast stations.	No explanation.	The submarine strata attract the plumb-line seawards.
	(7)	The inward deflections in longitude at Amritsar and Mooltan.	No explanation.	The Indo-Gangetic alluvium being over-compensated attracts the plumb-line away from the Himalayas.

The belt of negative maxima mentioned in the above summary is a remarkable feature of the table following page 14: an examination of this table discloses the fact, that between the parallels of 17° and 22°, whilst there is but *one* southerly deflection greater than 1″, there are 33 northerly deflections greater than 4″, 22 northerly deflections greater than 6″, 9 northerly deflections greater than 8″, and 3 northerly deflections greater than 10″. South of this belt, from latitude 17° to latitude 8°, no northerly deflection as large as 6″ has been as yet discovered.

Observed Deflections cannot be regarded as accidental.

Observed deflections cannot be treated as accidental for two reasons: *firstly*, the persistence of sign indicates the existence of a common cause, and *secondly*, small pockets of great density, sufficient to produce large northerly deflections at places to the south, should produce equally large southerly deflections at places to the north, and these reactions are only met with on rare occasions, *e.g.* at Kesri and Bangalore. The observers, who took the latitude observations on the meridian of 80° (*vide* table following page 14), and who from latitude 14° to 22° found a long succession of large negative values, ridicule the idea that each of these negative values is due to some separate local subterranean peculiarity: how is it they rightly ask, that we never meet with the opposite effects of these local hidden causes? Every visible hill, that we know of, which exercises a northerly attraction on a southern plumb-line, will exercise a southerly attraction on a northern plumb-line: and yet subterranean irregularities are readily assumed to account for inexplicable northerly deflections, and no corresponding opposite effect is demanded from them. If we examine the stations on the meridian of 74° from latitude 14° to latitude 24°, we find 13 negative values to 2 positive, both of the latter being less than 1″. On the meridian of 77° 30′ between latitude 15° and latitude 24° there is no positive value. Can it be maintained that these long successions of negative values are accidental?

The observed meridional deflections from south to north, if plotted graphically, degrees of latitude being measured on the axis of x, and seconds of deflection on the axis of y, form an undulating curve. From latitude 8° to latitude 14° the deflections are positive; in latitude 14° the curve cuts the axis of x and in latitude 20° attains a negative maximum (on all meridians): it again cuts the axis of x in latitude 24°, and after reaching a positive maximum in latitude 25° crosses the axis of x for the third time. The application of a negative correction to the fundamental latitude at Kaliánpur has the mere effect of lowering the axis of x with regard to the curve. The significance of the curve lies as much in its undulations as in its preference for the negative side of the axis; a negative correction applied to the fundamental latitude eliminates the latter peculiarity but leaves the former untouched.

No theory of partial compensation of visible mountains and seas will suffice to explain the preponderance of negative deflections, that have been observed in Central India, or the latitudinal belt of negative maxima crossing India in latitude 20°.

If we return to the summary of available evidence given in Table XVI (page 101), we find, speaking generally, that the latitude observations support the theory of "no compensation", and that the longitude observations support the theory of "complete compensation"; but this is not the whole difficulty, for some of the observed phenomena, *e.g.* the large deflections in Sub-Himalayan regions, admit of no explanation, if we adopt the theory of complete compensation, and other of the phenomena, *e.g.* the seaward deflections in longitude, cannot be explained, unless we do accept that theory. When confronted with a direct conflict of testimony we have to infer, that there is an unknown cause affecting our data, and rendering our conclusions abortive. In view of the fact that the deflections apparently change their direction on crossing the parallel of our station of origin, *a natural suspicion is that our adopted ellipsoid of reference is causing errors in our geodetic values.* Before however the question of the ellipsoid is considered, it is advisable to anticipate an argument that will assuredly occur to everyone: the theory of "entire compensation", it will be acknowledged, has been shown to be insufficient to explain phenomena, and the theory of "no compensation" has been found to be similarly insufficient; but it will be asked, cannot a compromise be effected, and will not a theory of "partial compensation" be found to satisfy all results? This question can only be answered in the negative: no hypothesis of a partial or irregular compensation of the Himalayas or Ocean is sufficient to account for observed phenomena.

In Part I of this paper I suggested that the alternations of negative and positive zones, the undulations of the curve, were possibly due to the joint attraction of the Himalayas and Ocean; for this idea to be upheld, the Ocean must be shown to be repelling the plumb-lines landwards and to be causing *northerly* deflections in the meridian. The longitude results in southern India admit of no other conclusion but that the compensation of the Ocean is entire; and we have to accept the inevitable corollary, that the Ocean is exerting no great influence on our plumb-lines in the meridian. When I attributed the alternation of zones to the joint effects of external mountains and seas, I did not foresee how unyielding to modification and explanation the longitude results would prove, and how impossible it was, in spite of their paucity and want of weight, to reject them.

The dominant feature of the latitude observations is the preponderance of the negative sign of (A − G) between the parallels of 14° and 24°: the negative sign denotes that the deflections in southern latitudes are more northerly than at Kalíánpur in latitude 24° 7′. *It is evident then that there is some powerful cause producing larger northerly deflections between latitude 14° and latitude 24° than between latitude 24° and latitude 25°.* What is this cause, if it be not the Ocean? The Himalayas, whether uncompensated or partially compensated, cause a greater northerly deflection at Kalíánpur than at stations south of Kalíánpur, and *therefore tend to render all values of (A − G) between the parallels of 14° and 24° positive*: if the Ocean actually *attracts* plumb-lines *towards* itself, as the longitude results lead us to believe, *it will also confer a positive tendency on values of (A − G) south of the parallel of 24°.* It is therefore clear that no theories of irregular and partial compensation suffice to explain the preponderance of negative values: the longitude results stubbornly impress upon us the initial necessity of admitting the entire compensation of the Ocean, and when this admission is once made, no theory of *partial* Himalayan compensation will account for the absence of positive values of (A − G) south of latitude 24°*.

---

* It is desirable to observe for Longitude at various points along the transcontinental arcs so that we may not be dependent on the results at the terminals: it would be well to extend the Amritsar-Mooltan Arc outwards to the hills, and to measure an arc in latitude 31° with one terminal near the Himalayas and the other at the foot of the Baluch plateau: the insertion of two longitude stations between Bombay and Waltair and between Mangalore and Madras would be of value.

TABLE XVII.

VALUES OF THE AXES OF THE EARTH.

	Date of determination	Major Axis in feet	Minor Axis in feet	Ellipticity
Laplace	1799	20919768	20852822	$\frac{1}{312 \cdot 20}$
Everest	1830	20922932	20853375	$\frac{1}{300 \cdot 80}$
Airy	1830	20923713	20853810	$\frac{1}{299 \cdot 33}$
Bessel	1841	20923600	20853656	$\frac{1}{299 \cdot 15}$
Clarke	1880	20926202	20854895	$\frac{1}{293 \cdot 47}$

In the following table are given the values of (A – G) in latitude, when the Clarke and Everest spheroids are used, and when Clarke's major axis is combined with Everest's value of the ellipticity, and when Everest's major axis is combined with Clarke's value of the ellipticity. Everest's account of his deduction of the Figure of the Earth shows considerable uncertainty, and we might now have been using either the third or fourth spheroid of the following table: there is therefore nothing extravagant in their conception, and it is interesting to see the values of (A – G) in latitude, with which we should have been confronted, had one of these spheroids been perchance adopted in 1830. The northern positive zone is represented in this table by Daiádhari, the southern positive zone by Punnæ, and the belt of negative maxima by Takalkhera.

TABLE XVIII.

Stations on the Great Arc of India, Meridian of 77½°	Latitude	Values of (A – G) in latitude			
		Everest's Spheroid a = 20922932 e = $\frac{1}{300 \cdot 80}$	Clarke's Spheroid 20926202 $\frac{1}{293 \cdot 47}$	Third Spheroid 20926202 $\frac{1}{300 \cdot 80}$	Fourth Spheroid 20922932 $\frac{1}{293 \cdot 47}$
	°	″	″	″	″
Dehra Dún ...	30	–38	–37	–34	–40
Kaliána ...	29	– 7	– 6	– 4	– 9
Daiádhari ...	25	+ 1	+ 1	+ 2	0
Kalíánpur ...	24	...	...	...	...
Takalkhera ...	21	– 7	– 7	– 9	– 6
Dámargída ...	18	– 3	– 3	– 6	0
Namthabad ...	15	– 1	– 2	– 6	+ 3
Punnæ ...	8	+ 2	+ 2	– 7	+ 8

The table shows

(i.) that a large deflection always appears at Dehra Dún,

(ii.) that the Clarke and Everest spheroids both give a positive zone in South India at Punnæ and another in Central India at Daiádhari,

(iii.) that on the third spheroid the positive zone at Daiádhari is accentuated, but the southern positive zone obliterated,

(iv.) that, if the third spheroid had been adopted no positive value of (A – G) would have now existed south of latitude 24°,

(v.) that on the fourth spheroid the northern positive zone is obliterated, and the southern positive zone largely enhanced and extended,

(vi.) *that the belt of negative maxima, crossing the Great Arc at Takalkhera, is perceptible on every spheroid.*

In Part I of this paper the question was considered, whether it would be correct to apply the mean Indian value of (A – G) in latitude as a correction to the fundamental latitude of India. The above table shows, that, if the fundamental latitude of India is to be corrected by a mean value of (A – G), the amount of the correction will depend on the spheroid adopted.

	Mean value of (A – G) omitting Himalayan stations
On the Everest spheroid	– 2″·5
,, Clarke ,, ...	– 2 ·5
,, third ,, ...	– 4 ·9
,, fourth ,, ...	– 0 ·7

A decrease of 1000 feet in the major axis of the fourth spheroid would suffice to render the mean value of (A – G) positive.

When we come to calculate the geodetic values of the longitude arcs on the third and fourth spheroids, we find that they are mainly dependent on the length of the major axis and are not sensibly affected by changes in the ellipticity.

TABLE XIX.

Arc	Latitude	Values of (A – G) in Longitude			
		Everest's Spheroid	Clarke's Spheroid	Third Spheroid	Fourth Spheroid
	°	″	″	″	″
Madras–Mangalore	13	– 9	– 6	– 6	– 9
Moulmein–Waltair	17	–15	– 6	– 6	–15
Waltair–Bombay	18	–10	– 4	– 4	–10
Chittagong–Karachi	24	–11	+ 4	+ 3	–11
Fyzabad–Quetta	28	+ 2	+11	+10	+ 2
Amritsar–Mooltan	31	+ 8	+10	+10	+ 8

If the Major axis is constant, the ellipticity may be changed from $\frac{1}{315}$ to $\frac{1}{288}$ without the values of (A — G) in longitude being appreciably altered. The seaward deflections at Madras and Mangalore can be eliminated by an increase of 6700 feet in Clarke's major axis, but such an increase will produce enormous positive values of (A — G) on the northern longitude arcs.

The seaward deflections cannot therefore be attributed to errors of the ellipsoid of reference, and *must be accepted as established and real.* Clarke's major axis is the most suitable for the Indian longitude arcs: any reduction of his value increases the (A — G) of southern arcs and any enlargement of his value increases the (A — G) of northern arcs.

Having been compelled to adopt Clarke's major axis to suit the Indian longitudes, we wish to learn, whether we can eliminate the belt of negative maxima of (A — G) in latitude and the alternations of latitudinal zones, and whether we can minimise the Sub-Himalayan deflections by introducing any particular value for the *ellipticity* of the spheroid.

TABLE XX.

Stations on the Meridian of 77½°	Values of (A — G) in latitude, employing Clarke's major axis.			
	Ellipticity = $\frac{1}{311 \cdot 04}$	$\frac{1}{300 \cdot 8}$	$\frac{1}{293 \cdot 47}$	$\frac{1}{289}$
	"	"	"	"
Dehra Dún	−31	−35	−37	−39
Kaliána	− 2	− 4	− 6	− 8
Daiádhari	+ 2	+ 1	+ 1	0
Kaliánpur	...	...	...	...
Takalkhera	−10	− 9	− 7	− 7
Dámargída	−10	− 6	− 3	− 2
Namthabad	−12	− 6	− 2	+ 1
Punnæ	−17	− 6	+ 2	+ 6

This table shows

(i.) that the belt of negative maxima, as represented by Takalkhera, can only be eliminated by the adoption of a value of the ellipticity considerably larger than $\frac{1}{289}$,

(ii.) that the reduction of large Sub-Himalayan deflections requires an ellipticity smaller than $\frac{1}{311}$,

(iii.) that a reduction of Sub-Himalayan deflections is necessarily accompanied by an enhancement of the preponderance of negative values in South India,

(iv.) that *vice versa* the elimination of the preponderance of southern negatives is accompanied by an increase of the Sub-Himalayan deflections,

(v.) that the alternation of zones cannot be eliminated by any modification of the spheroid.

It may therefore be concluded that our adopted spheroid is not a source of serious error, and that the Indian observed latitudes favour the Clarke spheroid.

The inferred existence of a hidden chain.

Two facts have been established on a fairly sound basis, *1stly*, that on the coast of India there is a slight deflection of the plumb-line towards the sea, and *2ndly*, that in Sub-Himalayan regions there is a marked deflection towards the Himalayas. The conclusions are that the Ocean is completely compensated, and that the Himalayas are not. Neither the assumption of a southerly local attraction at Kaliánpur nor the adoption of a modified ellipsoid of reference will eliminate the alternation of positive and negative zones; and the existence of an undiscovered cause of deflections has been inferred. It is now suggested that this undiscovered cause is perhaps a great invisible chain of excessive density, traversing India from Balasore near the mouth of the Hooghly to Jodhpur in Rajputana, and underlying Mandla and Bhopal. I have repeatedly deprecated assumptions of subterranean irregularities of density, unless their existence be confirmed by observations, showing, as at Moscow, *the two opposite effects of the hidden cause.* Such an objection however cannot be urged against the hypothesis of the subterranean chain described above, as in this case the two opposite effects are perceptible throughout. If we examine Chart No. 6, and the table following page 14, we shall find that between the parallels of 24° and 26° the plumb-lines are mostly deflected southwards, and that between the parallels of 21° and 18°, the deflections are northerly and large.

TABLE XXI.

On the Meridian of	the plumb-lines at	are deflected
73°	Jambo and Chaniána ...	14″ inwards
74½°	Jetgarh and Deo Dongri ...	6″ inwards
76°	Gurária and Kanheri ...	8″ inwards
77°	Daiádhari and Badgaon ...	9″ inwards
78°	Salímpur and Vánákonda ...	6″ inwards
80°	Pavia and Sítápár ...	10″ inwards
82°	Gurwáni and Patháidi ...	6″ inwards
84°	Huríláong and Khundábolo ...	16″ inwards
86°	Chendwár and Cuttack ...	12″ inwards
88°	Malúncha and Chandípur ...	4″ inwards

Throughout the entire length of the supposed chain the plumb-lines on either side of it are deflected towards each other. If this is the case, it may be asked, why was not such an obvious fact stated before? The fact is not obvious: the very magnitude of the chain has concealed its presence: it only becomes apparent after examination and analysis. That deflections of the plumb-line are due to *local* attractions is the accepted idea: every unexplained deflection is regularly ascribed to a cause hidden *in the immediate vicinity.* When we meet with a southerly deflection in latitude 27° and a northerly deflection in latitude 18°, we habitually prefer to assume two separate *local* causes to attributing both deflections to a distant central one.

If the chain had traversed southern India, its effects would have been perhaps more perceptible, but by an extraordinary coincidence *it crosses India in the latitude of the fundamental station of the Trigonometrical Survey.* When deflections change their sign at crossing the parallel of the station of reference, it is but natural to suspect either errors in the ellipsoid of reference, or the existence of great external sources of attraction; *not until we have shown that no modification of the ellipsoid of reference is admissible, are we justified in accepting the inward deflections along the chain as real: not until we have proved that visible external sources of attraction are insufficient to explain phenomena, are we warranted in assuming an invisible cause.**

The chain by itself does not suffice to explain all phenomena, but if we imagine the effects of a chain superimposed on those of a far-reaching Himalayan attraction, the alternations of zones, the undulations of the curve, the belt of negative maxima will become intelligible. On the Gangetic plain the attractive force of the chain is opposed to the attraction of the Himalayas, and though it may offer but a slight check to the latter, *it successfully obscures true Himalayan effects.* However large Himalayan attraction may be, its *variation* between stations in the plains 50 miles apart is small, and a chain capable of producing southerly deflections of 5″ would cause a positive zone, *i.e. apparent* southerly deflections at all those stations north of Kaliánpur, at which the effect of the Himalayas was within 5″ of its effect at Kaliánpur itself. South of the line Balasore–Mandla–Jodhpur the Himalayas and the chain combine forces, and after producing a belt of negative maxima, or large northerly deflections, they continue to exercise attraction as far as Cape Comorin. The chain does not *compensate* the attraction of the Himalayas: it imposes a wave on the latter's curve.

The Balasore–Mandla–Jodhpur line traverses the hilly region of Central India, and the question naturally arises, whether the visible ranges of hills, the Rajmahals, the Káimurs, the Vindhyas, the Satpuras, the Aravallis, the Mahádeo Pahár, are capable of producing the effects attributed to the hidden chain. An examination of maps can only lead to the one answer, that *the superincumbent masses of these mountains are insufficient to cause the observed deflections.*

On the meridian of 82° where the opposite and inward deflections are least apparent (see table following page 14) the mountains are most conspicuous. On the meridian of 80° where marked southerly deflections extend from latitude 23° 11′ to latitude 26° 54′, and marked northerly deflections from 22° 13′ to 18° 54′, the Mandla hills are comparatively insignificant.

On the meridian of 77° 30′ there is no apparent cause for the change in the sign of the deflections in latitude 24°: the average heights of compartments round Kaliánpur are given between pages 27 and 37, from which the general flatness of the country may be gauged. At Takalkhera a northerly deflection would certainly be expected, this station being situated only 20 miles south of the centre of the Mahádeo Pahár range: using a density-ratio of $\frac{1}{1 \cdot 67}$ (which is equivalent to assuming the Mahádeo Pahár rocks to have a density of 3·4), Colonel Everest calculated the northerly deflection at Takalkhera to be 5″: if we employ a density-ratio of ½, the calculated deflection will be 4″; the actual observed deflection is 6″·9. At Badgaon, 25 miles south of Takalkhera and 45 miles distant from the Mahádeo Pahár, the observed deflection is 7″·8. Even therefore in this neighbourhood the northerly deflections are not explained, that at Takalkhera being 2″·9 greater than would be expected, and that at Badgaon, which is more distant from the

---

* I acknowledge to have attached undue significance in the earlier parts of this paper to the fact that the deflections change their sign *along the parallel of the station of origin.* In Part I, I attributed the change of sign *along this line* to the fact, that it *was* the parallel of Kaliánpur. I now believe that the proximity of this line to the station of origin is an accidental coincidence.

mountains, being greater still*. At Ládi, *north* of the Mahádeo Pahár, the maps would lead one to expect a southerly attraction, but the deflection is still northerly, being 5″·3†.

On the meridian of 73° there is no visible cause of the great deflections at Deesa and Chaniána: the hill of Mount Abu might deflect the plumb-line at Deesa by half a second: the Aravalli hills might be expected to deflect the plumb-line at Chaniána by 3″, but it is out of the question to ascribe the actual deflections of 8″ and 11″ to the unaided force of Abu and the Aravallis.

It is instructive to study the views of the astronomical observers, who took the latitude observations in the northern positive zone and in the belt of negative maxima, and who at the time of observation recorded their opinions of the probable effects of visible mountain masses. In 1885 Colonel Heaviside wrote—"The country from Rámuápur to Pavia appears perfectly flat "* * * the southerly attraction shown at Etora, Dewarsán and Kánákhera (meridian of 80°) "is unaccounted for": Mr. Eccles recorded, that he saw no reason for the large northerly deflections, that he discovered at Díwai, Ankora and Burgpaili (meridian of 80°). Major-General Campbell did not anticipate a southerly deflection at Kesri, nor did Capt. Lenox Conyngham at Sironj (meridian of 77½°); Col. Herschel was not led from his study of the ground to expect a northerly deflection at Badgaon (meridian of 77½°), nor was Major-General Campbell at Ládi.

It may be concluded that the visible mountain ranges of Central India are of not sufficient mass to account for observed phenomena.

The effects attributed to the chain may be *partly* caused by the Indo-Gangetic alluvium, by the Deccan trap, and by the mountains of Central India, but they are *mainly* due to a *hidden* cause, that continues the line of inward deflections to the east and to the west, into Orissa and Rajputana, where the trap and the mountains have ceased to be visible.

The hidden chain appears to run parallel to the Himalayas.

The Longitude arcs Agra–Karachi, Agra–Deesa, Agra–Mooltan and Amritsar–Mooltan‡ (*vide* Chart No. 11) appear to traverse some invisible source of attraction, and lead one to imagine, that the chain is bending in the Rajputana desert to the north-west, and thus maintaining a strange parallelism with the Himalayas. The deflections shown by these four arcs, all of which traverse flat low-lying plains, are:—

Agra–Karachi, plumb-lines deflected	...	12″·77	inwards
Agra–Deesa	,, ...	12″·80	,,
Agra–Mooltan	,, ...	14″·95	,,
Amritsar–Mooltan	,, ...	10″·43	,,

Other arcs in this region show no such inward deflections:—

Mooltan–Karachi, plumb-lines deflected		2″·09	outwards
Deesa–Karachi	,, ...	0″·06	outwards
Deesa–Mooltan	,, ...	2″·09	inwards.

* "The (Badgaon) group," writes General Walker, "contains four stations whose distances range from 20 to 28 "miles to the south of Takalkhera, all which show a still larger amount of northerly attraction. Here therefore there "must be not only an excess of visible matter above ground in the Mahádeo plateau to the north, but a deficiency of in- "visible matter underground to the south." Philosoph. Transact. Royal Society, Vol. 186, p. 805.

† The Nerbudda Valley intervenes between Ládi and the Mahádeo Pahár, and tends to counteract the attraction of the latter on plumb-lines to the north: but the Nerbudda Valley is not a significant feature of land surface and should exercise no strong effect. The truth is that the interval between Ládi and Takalkhera is too great, and an intermediate latitude is wanted at the summit of the Mahádeo Pahár.

‡ The latitude and longitude of Agra, Karachi, Deesa, Amritsar and Mooltan are given in the table following page 14. Karachi is shown on Chart No. 7, and Deesa, Amritsar and Mooltan on Chart No. 12.

The data are insufficient to justify a definite assertion, that the inward deflections on the Agra–Karachi, Agra–Deesa, Agra–Mooltan and Amritsar-Mooltan arcs are due to a continuation of the same chain, as has been inferred to underlie Mandla, Bhopal and Jodhpur; nevertheless it is clear that these arcs of longitude reproduce in the Punjab and Rajputana the same extraordinary result, as has been given by the latitude observations in Bengal and the Central Provinces: south of the Himalayas the latitude observations have shown the existence of a hidden chain, that obscures Himalayan effects and deflects plumb-lines on either side of it towards itself: west of the Himalayas the Amritsar–Mooltan longitude observations disclose the existence of a hidden cause, deflecting plumb-lines towards itself, and also perhaps masking Himalayan effects. On Chart No. 12 I have marked with a cross all places, at which a hidden source of attraction *has been found to exist.* Whether the cross in the Punjab, as shown on the chart, belongs to the same chain as the crosses in Bengal, is a question, which cannot be answered, until more data have been accumulated. On Chart No. 12 the boundaries of the Tibetan plateau and of the Sub-Himalayan region have been plotted.

The effects of the hidden chain.

To disentangle the effects of the Himalayas from those of the chain and to obtain definite numerical ideas are operations complicated by the proximity of the chain to the station of origin: if as seems probable the chain crosses the meridian of 77° 30′ between Ládi and Ahmadpur, it would deflect the plumb-line at Kaliánpur to the south. Though reliable numerical results cannot be deduced from the data at present available, the following attempts to bring about an agreement between the results of observation and theory serve to illustrate the tendencies of the data. The ocean has been shown to be completely compensated and may be omitted from consideration: the uncompensated attraction of the Himalaya Mountains has been calculated to produce *approximately* the following deflections at stations of the Great Arc of India, (page 94, Table X)*.

TABLE XXII.

Stations† on the Meridian of 77°½	Regions of which the stations are types	Latitude	Calculated Northerly Deflections due to Himalayas	Calculated Differences from Punnæ
		°	″	″
Dehra Dún ...	Sub-Himalayan ...	30	72·2	69
Kaliána ...	... ...	29	36·2	33
Noh	Northern positive zone ...	*28*	*28*	25
Daiádhari ...	Ditto do.	*25*	*20*	17
Kaliánpur ...	... ...	24	18·4	15
Ládi	Belt of negative maxima	*23*	*17*	13
Badgaon ...	Ditto do.	*21*	*12*	10
Dámargída ...	... ...	18	10·0	7
Namthabad ...	... ...	*15*	7	4
Punnæ	... ...	8	3·4	0

* The omission from Table XXII of the attraction of Continental India (page 94) is questionable. At Kaliána Kaliánpur and Dámargída its effects are small: at Punnæ it produces a northerly deflection of 8″·7, but this effect should be excluded, because on the coast the deflections are slight and seaward, and the statement, that the Ocean is wholly compensated, implies that *the joint effect of ocean and land* at a station on the coast is slight. The doubtful point is the southerly deflection of 12″·6 at Dehra Dún: the inclusion of this deflection would decrease the calculated value of the resultant northerly deflection at Dehra Dún from 72″ to 60″. The difference in the calculated deflections at Dehra Dún and Kaliánpur would then be 42″ against 38″ as observed. So close an agreement between theory and observation would be a strong argument against the existence of any Himalayan compensation.

† Interpolated stations in italics.

The *observed* values of deflections, as given in the table following page 14, are not absolute but differential from Kaliánpur: as Kaliánpur is now suspected of being under the influence of the chain, it will be well to make Punnæ the station of reference.

TABLE XXIII.

Station	Observed differential values of deflections	
	With Kaliánpur as origin	With Punnæ as origin
	″	″
Dehra Dún ...	−38	−40
Kaliána ...	− 7	− 9
Noh ...	0	− 2
Daiádhari ...	+ 1	− 1
Kaliánpur ...	0	− 2
Ládi ...	− 5	− 7
Badgaon ...	− 8	−10
Dámargída ...	− 3	− 5
Namthabad ...	− 1	− 3
Punnæ ...	+ 2	0

The discrepancies between observed and calculated results, the Himalayas being assumed *uncompensated,* will be as follows:—

TABLE XXIV.

Stations on the meridian of 77° 30′	Latitude		Differential values of deflections with Punnæ as origin		
			As calculated *vide* Table XXII	As observed *vide* Table XXIII	Discrepancy
	°	′	″	″	″
Dehra Dún	30	19	−69	−40	−29
Kaliána ...	29	31	−33	− 9	−24
Noh ...	27	51	−25	− 2	−23
Daiádhari ...	24	38	−17	− 1	−16
Kaliánpur ...	24	7	−15	− 2	−13
Ládi ...	23	8	−13	− 7	− 6
Badgaon ...	20	44	−10	−10	0
Dámargída	18	3	− 7	− 5	− 2
Namthabad	15	6	− 4	− 3	− 1
Punnæ ...	8	9	0	0	0

The evidence of the existence of a compensation rests on the results at stations situated between the parallels of 23° and 30°: whilst the observations at these stations indicate considerable—almost entire—compensation, those at Dehra Dún show that the attraction of the visible mass is compensated only to the extent of $\frac{28}{69}$ths or approximately $\frac{2}{5}$ths. Too much weight should not be attached to the result at any one station; the elevation of the Himalayas may have been accompanied by great irregular compressions, and Dehra Dún may lie south of a region of abnormal density: but there is nothing to lead us to suppose that this is the case; the pendulum observations at Mussooree and Dehra Dún, far from supporting such an hypothesis, have been held to show, that *deficiencies* of matter underlie these Sub-Himalayan regions: and the latitude observations at Nojli, Sarkára, Sirsa and Rámuápur, stations fifty miles from the Himalayas, tend to indicate that the great deflection at Dehra Dún will not be found an isolated exception. The Himalayan curve on Chart No. 10 shows that, if the Himalayas exercise their full influence, the latter will be clear and unmistakeable in *differential* results at the extreme north of India only, and when we find an unmistakeable effect in differential results at the north of India, we cannot overlook its significance.

In the following table are shown the discrepancies that occur between theory and observation, if Himalayan attraction is assumed to be compensated to the extent of one-third, one-half and two-thirds its normal force respectively.

TABLE XXV.

Stations	Discrepancies between theory and observation, if Himalayan attraction is assumed compensated to the extent of		
	$\frac{1}{3}$rd.	$\frac{1}{2}$	$\frac{2}{3}$rds.
	″	″	″
Dehra Dún	− 6	+ 6	+17
Kaliána	−13	− 7	− 2
Noh	−15	−10	− 6
Daiádhari	−10	− 7	− 5
Kalíánpur	− 8	− 5	− 3
Ládi	− 2	+ 1	+ 3
Badgaon	+ 3	+ 5	+ 7
Dámargída	0	+ 2	+ 3
Namthabad	0	+ 1	+ 2
Punnæ	...	...	...

It will be seen that discrepancies of 13″ and 15″ continue to appear if the theory of a one-third compensation be adopted. The theory of a two-thirds compensation renders the effects of the chain approximately equal and opposite at Kalíánpur and Ládi, and also at Noh and Badgaon*, but it introduces a discrepancy of 17″ at Dehra Dún.

---

* I am assuming the discrepancies between theory and observation to be due to the influence of the chain.

The expression "Himalayan Attraction" has hitherto in this paper been assumed to comprehend both the attraction of the Himalaya Mountains and that of the Tibetan plateau. If the uncompensated effects of this Himalayan attraction be analysed, and those due to the Himalayan ranges separated from those due to the Tibetan plateau, the following results are obtained:—

TABLE XXVI.

CALCULATED DEFLECTIONS.

Stations			Deflections due to		Total
			Himalayan Ranges	Tibetan Plateau	
			"	"	"
Dehra Dún	...	...	19	53·2	72·2
Kaliána	...	...	3	33·2	36·2
Noh	...	...	2	26	28
Daiádhari	...	...	0·6	19	19·6
Kaliánpur	...	...	0·4	18	18·4
Ládi	...	...	...	16	16
Badgaon	...	...	...	13	13
Dámargída	...	...	...	10·0	10·0
Namthabad	...	...	...	7	7
Punnæ	...	...	...	3·4	3·4

If it be assumed that the Himalayan ranges are uncompensated, and that the Tibetan plateau is partially compensated, the following discrepancies will remain between the results of calculation and observation.

TABLE XXVII.

Station		Discrepancies between calculation and observation, the attraction of the Tibetan plateau being assumed compensated to the extent of		
		$\frac{1}{3}$rd.	$\frac{1}{2}$	$\frac{2}{3}$rds.
		"	"	"
Dehra Dún	...	−12	− 4	+ 4
Kaliána	...	−14	− 9	− 4
Noh	...	−15	−11	− 8
Daiádhari	...	−10	− 8	− 5
Kalíánpur	...	− 8	− 6	− 3
Ládi	...	− 2	+ 1	+ 3
Badgaon	...	+ 4	+ 5	+ 7
Dámargída	...	+ 1	+ 2	+ 3
Namthabad	...	+ 1	+ 1	+ 2
Punnæ	...	0	0	0

The hypothesis, that the attraction of the Hímalayan ranges is uncompensated, and that that of the Tibetan plateau is compensated to the extent of two-thirds its normal force, removes all gross discrepancies, and renders the effects of the chain at stations to the north and south of it approximately equal and opposite. It makes the crest of the chain cut the meridian of 77° 30′ in latitude 23° 30′, and it attributes to the chain a maximum effect of 8″ at 200 miles north and south of its crest.

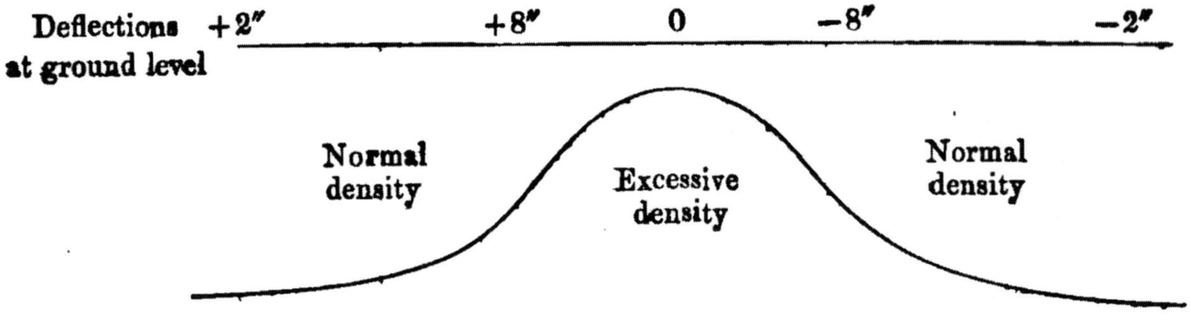

In the following table the possible effects of the chain are illustrated on the hypothesis that the attraction of the Tibetan plateau is compensated to the extent of two-thirds its normal force:—

TABLE XXVIII.

Station	Latitude	Calculated deflections due to			Total deflections by theory	Differential deflections from Punnæ		Discrepancy
		Himalayan Ranges	Tibetan Plateau	Underground chain		By theory	By observation	
	° ′	″	″	″	″	″	″	″
Dehra Dún ...	30 19	−19	−18	+1	−36	−35	−40	+ 5
Kaliána ...	29 31	− 3	−11	+2	−12	−11	− 9	− 2
Noh ...	27 51	− 2	− 9	+7	− 4	− 3	− 2	− 1
Daiádhari ...	24 38	− 1	− 6	+4	− 3	− 2	− 1	− 1
Kaliánpur ...	24 7	− 0·4	− 6	+3	− 3	− 2	− 2	0
Ládi ...	23 8	...	− 5	−3	− 8	− 7	− 7	0
Badgaon ...	20 44	...	− 4	−7	−11	−10	−10	0
Dámargída ...	18 3	...	− 3	−2	− 5	− 4	− 5	+ 1
Namthabad ...	15 6	...	− 2	−1	− 3	− 2	− 3	+ 1
Punnæ ...	8 9	...	− 1	0	− 1	...	...	...

This attempt to disentangle the effects of the chain from those of the Himalayas—the effects of the obscuring cause from those of the visible and compensating causes—is put forward as an illustration of the data and not as a solution of the problem. The point for which I contend is the recognition of the possible existence of an underground chain in Central India, and of the consequent obscuration of true Himalayan effects: this paper seemed incomplete without a reference to numerical results, but no stress can be laid on the latter*. If the northern slope of the chain be assumed almost vertical, and the southern slope almost horizontal, the opposite effects of its attraction will not be equal at places equidistant from its ridge: such a chain will create a positive zone north of latitude 24°, but its negative effects to the south, though perceptible to a greater distance, will be less marked: the gradual decrease in the *northerly* deflection of the plumb-line, that has been *observed* to continue through 800 miles, from Badgaon to Punnæ, from the belt of negative maxima to the southern positive zone, (Table XXIII, page 111) will then be mainly a Himalayan effect.

If we regard the Indo-Gangetic alluvium as an ocean, we find that its southern shore overlies the position of the chain: and we see that plumb-lines on this sea of sand and plumb-lines on the main land to the south are deflected towards the coast area, in which the change of elevation from the lower strata of the alluvium to the summits of the Vindhyas occurs: on the north of the sea of alluvium the deflections are towards the area, within which the Himalayas rise from the level of the sea to the level of the snow: on the eastern, southern and western coasts of India the plumb-lines hang outwards towards that area, within which the submerged cliffs drop from the level of the shore to the depths of the ocean. If then we can say, that the areas in which great changes of elevation occur are generally sources of attraction, we reconcile the contradictory effects of mountains and seas.

Dehra Dún:

December 1901.

---

* The numerical results are based on an unwarranted assumption, *viz.*, that the northern and southern slopes of the underground chain are inclined at the same angle to the vertical; if, as is probable, the cross-section of the chain is irregular in outline, the deflections will not be equal at equal distances north and south of its crest.

APPENDICES.

APPENDIX I.

The following is Captain Lenox Conyngham's description of the ground and country round Kaliánpur:—

Kaliánpur Group.

The main feature of the tract of the country in which the stations of the Kaliánpur group lie is the plateau, about 200 feet higher than the surrounding country, which, spreading widely to the north, narrows down to a wedge and ceases a little to the south of Kámkhera. The average height of the plateau above mean sea-level is about 1,700 feet.

The stations of Daiádhari, Súrantál, Kaliánpur, Losalli, Tinsia and Kámkhera are on the plateau. The N. E. End of the Base and Bhaorása are in the plain to the east and Ahmadpur on an isolated hill which rises out of the plain to the south about 15 miles from the southern end of the plateau. At Daiádhari the width of the plateau is so great that it appears to be merely a level plain. Súrantál and Kaliánpur are on the eastern edge, Tinsia on the western, Kámkhera on the southern, while Losalli is in the middle.

The immediate surroundings of the stations are as follows:—

Daiádhari. Height above mean sea-level 1,867 feet. The station stands on the highest point of an unimportant isolated hill about 100 feet above the surrounding plain. The top is fairly level and about 50 yards wide by 150 long, the direction of the length being north and south. The station is at the northern edge. Similar small hills are scattered over the country at intervals, the nearest being about 1½ miles to the north; with the exception of these small hills the country is flat.

Bhaorása. Is situated on the highest point of a small rolling hillock of sand-stone, 1,387 feet above mean sea-level, which rises very gently from the general level of the plain to the south and west but falls more abruptly to the north and east. The Betwa river runs by the eastern end of the hill at a distance of about 1½ miles from the station. The height of the station above the general level of the plain is under 100 feet.

N. E. End of the Base. The station lies in the plain to the east of Kaliánpur and is 1,481 feet above mean sea-level. The plain is perfectly flat and the horizon almost unbroken except to the west where the edge of the plateau rises slightly above it.

Kaliánpur. The station, which is 1,765 feet above mean sea-level, is on the highest of a series of rolling hills or downs which form the eastern edge of an extensive plateau about 170 feet higher than the plain to the east. The edge of the plateau runs north and then north-east and disappears in the distance; it is somewhat higher than the central parts and more undulating. The town of Sironj lies about 2½ miles to the south-east.

Súrantál. The situation of the station is very similar to that of Kaliánpur. It is 1,802 feet above mean sea-level. It is from a point very near this station that the edge of the plateau bends towards the east.

Losalli. Is situated 1,749 feet above mean sea-level and in the middle of the plateau in perfectly flat ground, which is slightly lower than the undulating country to the west.

Tinsia. Is situated very similarly to Kaliánpur but on the western edge of the plateau. It is 1,776 feet above mean sea-level. The station is surrounded for miles by dense jungle, but is not far from a track which runs from Sironj to the valley of the Parbatti.

Kámkhera. Is on a flat-topped hill near the southern end of the plateau. Its height is 1,780 feet above mean sea-level.

Ahmadpur. Is 1,715 feet above mean sea-level and is situated on a conspicuous hill of almost solid rock, which rises to a height of over 200 feet out of the low plain to the south of the Kaliánpur plateau. The ascent from the east is easy but on the other sides somewhat precipitous; there are many similar hills at intervals on every side but none so large. The nearest is a small one about 2 miles to the south-east. The plain between Kámkhera and Ahmadpur is about 1,430 feet above mean sea-level.

APPENDIX II.

RESULTS OF THE AZIMUTHS OBSERVED IN INDIA AND BURMA.

NOTE.—The computed Azimuths are based on General Walker's derived Azimuth of Súrantál at Kaliánpur, *viz.*, 190° 27′ 5″ 10. The last two columns show the combined effect of local attraction in the Prime Vertical and of the azimuthal error accumulated in the Principal Triangulation, *vide* page 19.

Section	Series	Station		Date of Observation	Latitude North = φ		Longitude East of Greenwich		Height above Sea Level	Observed *minus* Computed Azimuth = (A − G)	(A − G) × cot φ
					°	′	°	′	*feet*	″	″
N.-W. QUADRILATERAL	Karachi Longitudinal	Losalli	S.	January 1849 ...	24	6	77	36	1749	+ 1·37	W. 3·06
		Salot	H.S.	March 1849 ...	24	15	77	17	1834	− 0·73	E. 1·62
		Mátá-ká-húra	„	April 1849 ...	24	14	76	39	1645	+ 0·07	W. 0·16
		Gurária	„	Nov. and Dec. 1849 ...	24	26	76	7	1360	− 0·11	E. 0·24
		Rámpura	„	November 1849 ...	24	29	75	29	1920	0·00	0·00
		Aramlia	S.	February 1850 ...	24	25	75	2	1532	+ 1·72	W. 3·79
		Sánd	H.S.	February 1850 ...	24	43	74	35	1910	+ 2·90	„ 6·30
		Tiki	„	January 1851 ...	24	56	73	53	2369	+ 2·61	„ 5·61
		Kánnagar	„	December 1850 ...	24	58	73	21	3607	− 4·17	E. 8·96
		Gúru Sikkar	„	November 1850 ...	24	39	72	49	5650	+ 0·96	W. 2·09
		Birona	S.	November 1851 ...	24	27	72	16	673	− 1·65	E. 3·63
		Khankharia	„	March and April 1851	24	37	71	56	362	− 1·87	„ 4·08
		Sarla	„	November 1851 ...	24	47	71	37	132	+ 2·86	W. 6·19
		Didáwa	H.S.	December 1851 ...	24	51	71	21	212	+ 1·16	„ 2·50
		Virária	„	December 1851 ...	24	57	71	5	460	+ 1·76	„ 3·78
		Lúnki	„	December 1851 ...	24	58	70	42	588	+ 1·44	„ 3·09
		Rojhra	„	December 1851 ...	24	57	70	17	518	+ 0·05	„ 0·11
		Chánga	„	January 1852 ...	24	59	69	54	349	− 3·72	E. 7·98
		Mairáb-ka-Shahar	T.S.	January 1852 ...	24	50	69	23	44	− 0·02	„ 0·04
		Khori	„	February 1852 ...	25	1	69	6	63	− 1·53	„ 3·28
		Alamkhán	„	December 1852 ...	24	50	68	46	67	+ 2·07	W. 4·47
		Chútli	„	January 1853 ...	24	46	68	26	72	+ 2·69	„ 5·81
		Károthol	H.S.	February 1853 ...	24	54	67	56	260	+ 0·14	„ 0·30

Section	Series	Station	Date of Observation	Latitude North = φ	Longitude East of Greenwich	Height above Sea Level	Observed *minus* Computed Azimuth = (A − G)	(A − G) × cot φ
				° ′	° ′	*feet*	″	″
N.-W. QUADRILATERAL—(*Continued*).	N. W. Himalaya	Medwáni H.S.	Jan. and Feb. 1853 ...	31 18	76 14	1935	− 6·08	E. 10·00
		Jáoli „	December 1851 ...	33 17	73 13	1918	+ 0·01	W. 0·02
	Great Indus ...	Karachi Observatory	October 1855 ...	24 50	67 4	35	− 1·65	E. 3·57
		Karachi Base-line, S. End S.	March 1853 ...	24 53	67 12	69	− 1·46	„ 3·15
		Yúsuf P.S.	December 1858 ...	27 51	68 29	215	+ 1·26	W. 2·38
		Bhanar T.S.	April 1859 ...	28 9	69 20	256	+ 5·89	„ 11·01
		Míáni „	December 1859 ...	28 34	69 53	300	+ 9·96	„ 18·29
		Dájil P.S.	April 1860 ...	29 33	70 25	412	+ 11·85	„ 20·90
		Dera Dín Panáh „	April 1859 ...	30 34	70 59	490	+ 5·68	„ 9·62
		Jharkil T.S.	Dec. 1858 & Jan. 1859	31 21	71 2	554	+ 4·11	„ 6·75
	Great Arc Meridional (Section 24° to 30°)	Kalíánpur H.S.*	Dec. 1836 & Jan. 1837	24 7	77 42	1765	...	...
		Pahárgarh „	Dec. 1836 & Jan. 1837	24 56	77 44	1641	+ 1·89	W. 4·07
		Kesri „	December 1836 ...	25 47	77 43	1487	− 1·70	E. 3·52
		Usira „	February 1838 ...	26 57	77 40	810	+ 0·80	W. 1·57
		Noh T.S.	April 1837 ...	27 51	77 41	710	+ 2·47	„ 4·67
		Datairi „	January 1836 ...	28 44	77 41	767	− 0·56	E. 1·02
		Kaliána S.	October 1836 ...	29 31	77 42	828	− 1·22	„ 2·15
		Banog H.S.	September 1836 ...	30 29	78 3	7433	− 14·54	„ 24·70
		Dehra Dún Observatory (old)	March and April 1853	30 20	78 6	2289	− 12·18	„ 20·82
	Rahún Meridional ...	Kánkra H.S.	March and April 1862	25 38	76 10	1652	+ 1·09	W. 2·27
		Bánskho „	April 1861 ...	26 50	76 11	1870	+ 2·91	„ 5·75
		Tásíng „	December 1860 ...	27 53	76 15	2050	+ 4·09	„ 7·73
		Rákhi T.S.	December 1856 ...	29 17	76 9	785	+ 1·78	„ 3·17
		Kheri „	Jan. and Feb. 1856 ...	30 5	76 8	822	− 1·61	E. 2·78
		Bowra „	April 1853 ...	30 21	76 9	855	+ 1·74	W. 2·97

* Initial Azimuth Station.

Section	Series	Station		Date of Observation	Latitude North = φ		Longitude East of Greenwich		Height above Sea Level	Observed *minus* Computed Azimuth = (A − G)	(A − G) × cot φ
					°	′	°	′	*feet*	″	″
N. W. QUADRILATERAL—(*Continued*).	Gurhágarh Meridional	Rájgarh	H.S.	March 1863 ...	26	18	74	38	2618	+ 0·22	W. 0·45
		Garinda	S.	March 1863 ...	27	56	75	4	1204	+ 2·10	„ 3·96
		Sirsa	„	April 1861 ...	29	32	75	4	738	+ 1·11	„ 1·96
		Sangatpur	T.S.	March and April 1860	31	18	75	5	779	+ 2·63	„ 4·33
	Jogi-Tíla Meridional	Akbar	P.S.	January 1857 ...	30	54	73	20	641	− 0·84	E. 1·40
	Sutlej ...	Paphra	T.S.	March and April 1861	29	6	70	52	341	+ 3·81	W. 6·85
		Ládimsir	„	January 1862 ...	29	22	72	2	468	− 0·31	E. 0·55
		Mandresa	„	March and April 1862	29	55	73	2	512	+ 0·06	W. 0·10
		Jhambhera	„	December 1862 ...	30	6	73	52	630	− 3·63	E. 6·26
	Jodhpore Meridional	Thob	H.S.	March 1873 ...	26	3	72	25	856	+ 3·74	W. 7·65
		Jambo	„	Feb. and March 1874	27	16	72	34	772	− 0·65	E. 1·26
		Mugrala	„	February 1875 ...	28	31	72	25	517	− 2·13	„ 3·92
	Eastern Sind Meridional	Malar	H.S.	January 1877 ...	26	2	70	6	328	− 2·86	E. 5·86
		Asu	„	February 1880 ...	27	11	70	13	479	− 0·89	„ 1·73
		Vijnot	T.S.	Dec. 1880 & Jan. 1881	28	2	69	53	276	+ 4·04	W. 7·59
		Dáowála	„	February 1881 ...	28	20	69	53	282	+ 5·01	„ 9·29
S. E. QUADRILATERAL	Great Arc Meridional (Section 18° to 24°)	Ahmadpur	H.S.	December 1838 ...	23	36	77	43	1713	+ 0·82	W. 1·88
		Bhimbat	„	December 1838 ...	22	50	77	40	2120	+ 1·54	„ 3·66
		Nílgarh	„	February 1839 ...	21	46	77	42	2533	− 1·14	E. 2·86
		Badgaon	„	January 1839 ...	20	44	77	39	1128	+ 0·63	W. 1·66
		Sákri	„	December 1838 ...	20	0	77	45	1810	+ 1·41	„ 3·87
		Somtána	„	April 1838 ...	19	5	77	42	1714	− 2·36	E. 6·82
		Dámargída	S.	October 1838 ...	18	3	77	43	1946	− 1·89	„ 5·80

Section	Series	Station	Date of Observation	Latitude North = φ		Longitude East of Greenwich		Height above Sea Level	Observed minus Computed Azimuth = (A − G)	(A − G) × cot φ
				°	′	°	′	feet	″	″
S. E. QUADRILATERAL—(*Continued*).	Calcutta Longitudinal	Budhon H.S.	March 1864 ...	24	5	78	34	1867	− 0·29	E. 0·65
		Rangír H.S. (old)	January 1834 ...	24	0	79	28	1180	− 14·29	„ 32·10
		Amua H.S.	January 1834 ...	24	0	80	32	2113	+ 0·68	W. 1·53
		Karára „	April 1842 ...	24	5	81	18	1966	− 8·37	E. 18·73
		Gurwáni „	December 1845 ...	24	1	82	20	2083	+ 1·90	W. 4·26
		Gora „	December 1845 ...	24	5	83	17	1828	− 6·61	E. 14·79
		Huríláong „	Dec. 1848 & Jan. 1849	24	2	84	24	1378	− 9·10	„ 20·41
		Chendwár H.S. (old)	December 1843 ...	23	57	85	29	2820	− 4·08	„ 9·19
		Párasnáth H.S.	December 1850 ...	23	58	86	11	4481	− 5·99	„ 13·48
		Tilabani „	December 1845 ...	23	25	86	36	1329	− 3·73	„ 8·61
		Malúncha „	April 1844 ...	23	54	87	8	970	− 8·06	„ 18·19
		Madhpur T.S.	December 1868 ...	23	10	87	47	180	− 1·59	„ 3·72
		Aknápur „	March 1869 ...	22	54	88	6	98	− 6·70	„ 15·86
		Calcutta Base-line, S. End T.S.	Dec. 1844 & Jan. 1845	22	37	88	25	13	− 9·25	„ 22·20
	East Coast ...	Patna T.S.	April 1852 ...	21	47	87	14	80	− 7·44	E. 18·62
		Chandípur „	December 1854 ...	21	27	87	5	51	− 4·96	„ 12·62
		Cuttack H.S.	October 1854 ...	20	29	85	54	132	− 3·39	„ 9·08
		Khundábolo „	January 1857 ...	19	51	85	1	3115	− 4·91	„ 13·60
		Ráwal „	Dec. 1859 & Jan. 1860	18	32	83	36	874	− 2·18	„ 6·50
		Vizagapatam Base-line, N. End S.	Jan. and Feb. 1863 ...	18	1	83	16	181	− 1·93	„ 5·93
	Bider Longitudinal	Pirmulo H.S.	February 1869 ...	17	53	78	38	2093	− 2·50	E. 7·75
		Vánákonda „	Feb. and March 1869	17	36	79	25	1664	− 2·30	„ 7·25
		Singáwáram „	February 1871 ...	17	45	80	59	714	− 3·22	„ 10·06
		Kálingkonda „	January 1872 ...	17	50	82	21	4634	− 1·38	„ 4·29
		Sánjib „	December 1860 ...	17	31	82	44	2142	− 1·57	„ 4·97

Section	Series	Station	Date of Observation	Latitude North = φ	Longitude East of Greenwich	Height above Sea Level	Observed *minus* Computed Azimuth = (A − G)	(A − G) × cot φ
				° ′	° ′	*feet*	″	″
S. E. QUADRILATERAL—(*Continued*).	Jabalpur Meridional	Karaundi H.S.	Jan. and Feb. 1865 ...	23 11	80 2	1625	− 3·91	E. 9·13
		Sarandi Pat „	March and April 1865	22 13	80 6	1627	− 1·19	„ 2·91
		Bhímsain „	December 1866 ...	20 58	79 49	1490	− 1·02	„ 2·66
		Díwai „	January 1867 ...	19 50	79 35	967	− 2·42	„ 6·71
		Burgpaili „	February 1867 ...	18 54	79 44	983	− 2·65	„ 7·74
	Biláspur Meridional	Patháídi T.S.	December 1871 ...	21 49	82 19	879	− 2·46	E. 6·15
		Ramai H.S.	December 1872 ...	20 57	82 11	1313	− 2·49	„ 6·50
		Karía „	January 1873 ...	19 12	82 10	2014	− 2·28	„ 6·55
	South Malúncha Meridional	Kalaíbhánga T.S.	December 1849 ...	22 20	87 11	303	− 2·20	E. 5·36
N. E. QUADRILATERAL	North-East Longitudinal	Kalíánpur T.S.	March and April 1850	28 35	79 47	629	− 1·38	E. 2·53
		Rámuápur T. S. (old)	December 1838 ...	28 23	80 31	546	− 0·17	„ 0·31
		Mási T.S.	Dec. 1849 & Jan. 1850	27 38	81 26	426	− 5·80	„ 11·08
		Bansídíla „	April 1849 ...	27 24	82 19	377	− 4·08	„ 7·87
		Naonangarhi S.	June 1852 ...	26 59	84 26	344	− 7·36	„ 14·46
		Chúni T.S.	December 1846 ...	26 11	87 5	197	− 9·05	„ 18·41
		Rámganj „	Dec. 1852 & Jan. 1853	26 19	88 20	249	− 10·16	„ 20·54
	Budhon Meridional	Gúrmi T.S.	December 1842 ...	26 36	78 33	575	− 1·51	E. 3·02
		Sankráo „	February 1843 ...	28 2	78 35	670	+ 1·40	W. 2·63
		Sirsa „	February 1843 ...	28 55	78 35	739	− 4·23	E. 7·66
	Rangír Meridional	Muhammadabad T.S.	December 1840 ...	27 18	79 28	565	+ 7·67	W. 14·86
	Amua Meridional	Nimkár T.S.	April 1838 ...	27 21	80 32	486	+ 4·65	W. 8·99
	Karára Meridional	Pabhosa H.S.	June and July 1845 ...	25 21	81 22	565	− 3·27	E. 6·90
		Sora T.S.	October 1845 ...	26 17	81 15	409	+ 4·50	W. 9·11

Section	Series	Station		Date of Observation	Latitude North = φ		Longitude East of Greenwich		Height above Sea Level	Observed *minus* Computed Azimuth = (A − G)	(A − G) × cot φ
					°	′	°	′	*feet*	″	″
N. E. QUADRILATERAL—(*Continued*).	Gurwáni Meridional	Marár	T.S.	April 1846 ...	25	41	82	17	371	− 3·91	E. 8·13
		Bisaul	„	Jan. and Feb. 1847 ...	26	41	82	23	342	− 4·30	„ 8·56
	Gora Meridional	Hirdepur	T.S.	March and April 1846	25	24	83	17	289	− 4·03	E. 8·49
		Samenda	„	December 1846 ...	26	0	83	16	285	− 2·22	„ 4·55
		Rájábári	„	April 1847 ...	26	54	83	18	296	− 4·32	„ 8·52
	Hurïláong Meridional	Mednipur	T.S.	February 1850 ...	25	5	84	25	335	− 6·74	E. 14·40
		Jalálpur	„	February 1852 ...	26	4	84	26	232	− 1·38	„ 2·82
	Chendwár Meridional	Pota	T.S.	April 1846 ...	26	23	85	29	222	− 6·38	E. 12·86
	North Párasnáth Meridional	Bichwi	H.S.	December 1851 ...	25	10	86	11	321	− 6·27	E. 13·34
	North Malúncha Meridional	Sirkanda	T.S.	April 1846 ...	25	28	87	11	132	− 6·62	E. 13·90
	Calcutta Meridional	Anandbás	T.S.	Dec. 1845 & Jan. 1846	23	21	88	25	67	− 7·96	E. 18·44
		Madhupur	„	December 1846 ...	23	57	88	32	92	− 9·34	„ 21·03
	East Calcutta Longitudinal	Daulatpur	T.S.	December 1868 ...	23	9	89	45	60	− 4·67	E. 10·92
		Gangapur	„	April 1866 ...	23	0	90	30	54	− 6·88	„ 16·21
		Lakhinagar	„	December 1866 ...	23	1	90	48	51	− 1·66	„ 3·91
	Brahmaputra Meridional	Tepri	T.S.	December 1869 ...	23	57	89	55	67	− 7·40	E. 16·66
		Aloákándi	„	March 1873 ...	24	45	89	41	88	− 8·62	„ 18·70
		Halkáchar	„	April 1873 ...	25	10	89	45	103	− 12·27	„ 26·11
	Eastern Frontier (Section 23° to 26°)	Rangsanobo	H.S.	Oct. and Nov. 1861 ...	25	15	91	46	4455	− 9·61	E. 20·38
		Dawa	„	Dec. 1863 & Jan. 1864	23	45	91	23	205	− 7·28	„ 16·55

Section	Series	Station	Date of Observation	Latitude North = φ	Longitude East of Greenwich	Height above Sea Level	Observed *minus* Computed Azimuth =(A − G)	(A − G) × cot φ
				° ′	° ′	*feet*	″	″
N. E. QUADL. (*Continued*).	Assam Longitudinal	Ataro Bánki T.S.	Dec. 1855 & Jan. 1856	26 5	89 31	133	− 11·14	E. 22·76
		Alangjáni „	February 1874 ...	25 59	89 48	143	− 10·78	„ 22·12
		Raikusni H.S.	November 1858 ...	26 8	90 42	803	− 11·97	„ 24·40
SOUTHERN TRIGON	Great Arc Meridional (Section 8° to 18°)	Kodangal S.	January 1872 ...	17 8	77 41	1906	− 4·00	E. 12·98
		Darúr H.S.	March 1871 ...	16 14	77 42	1796	− 4·99	„ 17·14
		Bangalore Base-line, S.W. End S.	May 1870 ...	13 1	77 37	3126	− 5·67	„ 24·53
		Bangalore Base-line, N.E End S.	May 1870 ...	13 5	77 42	3016	− 5·44	„ 23·41
		Kanjamalai H.S.	Nov. & Dec. 1869 ...	11 37	78 6	3236	− 7·57	„ 36·82
		Pachapálaiyam Station	February 1870 ...	11 0	77 40	970	− 5·80	„ 29·84
		Kutipárai S.	December 1873 ...	9 29	78 3	351	− 8·12	„ 48·61
		Rádhápuram „	March 1869 ...	8 17	77 45	170	− 6·74	„ 46·30
		Kudankulam „	March 1869 ...	8 10	77 44	177	− 8·48	„ 59·09
	Bombay Longitudinal	Achola H.S.	December 1840 ...	18 15	77 2	2274	− 2·72	E. 8·25
		Nitali „	November 1840 ...	18 17	76 19	2289	− 6·37	„ 19·28
		Kanheri „	December 1837 ...	18 30	75 46	2610	− 4·69	„ 14·02
		Alsunda „	March 1863 ...	18 27	75 3	2163	− 5·41	„ 16·22
		Khánpisura „	October 1846 ...	18 46	74 49	2751	− 6·66	„ 19·60
		Dhauleshvar „	April 1838 ...	18 26	74 12	2939	− 3·97	„ 11·91
		Mándvi „	March 1841 ...	18 38	73 35	4121	− 6·37	„ 18·89
	South Konkan Coast	Mirya H.S.	October 1844 ...	17 2	73 18	473	+ 1·26	W. 4·11
		Chaukola „	December 1848 ...	15 56	74 2	2794	− 2·65	E. 9·28
		Kumbhári „	January 1844 ...	15 9	74 20	2898	+ 0·09	W. 0·33
	Mangalore Meridional	Páchvad H.S.	March 1865 ...	17 31	74 42	3138	− 5·25	E. 16·63
		Karabgati „	December 1865 ...	16 8	74 50	2544	− 5·30	„ 18·32
		Koramúr „	March 1878 ...	14 8	75 1	2525	− 6·76	„ 26·85

Section	Series	Station	Date of Observation	Latitude North = φ		Longitude East of Greenwich		Height above Sea Level	Observed minus Computed Azimuth = (A − G)	(A − G) × cot φ
				°	′	°	′	feet	″	″
SOUTHERN TRIGON—(Continued).	Madras Meridional and Coast	Dhúlipalla S.	April and May 1868...	16	26	80	8	244	− 4·83	E. 16·38
		Dánapa H.S.	December 1863 ...	15	56	79	59	1010	− 6·04	„ 21·16
		Kistama „	December 1864 ...	14	27	79	48	458	− 4·50	„ 17·46
		Parampúdi „	December 1861 ...	17	13	81	15	685	− 6·23	„ 20·11
	South-East Coast and Ceylon Branch	Kallapat Trs. S.	March 1879 ...	11	57	79	36	199	− 5·15	E. 24·33
		Nayinipiriyán „	January 1879 ...	11	8	79	23	158	− 2·17	„ 11·03
		Pátharankota „	March 1877 ...	10	28	79	15	120	− 7·69	„ 41·63
		Manĕgandi „	February 1876 ...	9	46	78	58	56	− 9·41	„ 54·67
		Ramnad S.	March 1875 ...	9	22	78	52	48	− 7·97	„ 48·32
	Madras Longitudinal	Mangalore S.	March 1878 ...	12	52	74	53	185	− 3·10	E. 13·57
		Nughallibĕtta H.S.	November 1871 ...	13	2	76	31	3140	− 7·64	„ 33·00
		Anandalamalai „	January 1866 ...	12	56	79	26	923	− 5·48	„ 23·86
		St. Thomas's Mount Trs. S.	February 1880 ...	13	0	80	14	250	− 4·84	„ 20·96
		Injambákam H.S.	February 1880 ...	12	55	80	18	29	− 5·12	„ 22·33
S. W. QUADRILATERAL	Khánpisura Meridional	Indráwan T.S.	March and April 1847	22	49	75	13	1834	− 0·44	E. 1·05
		Valvádi H.S.	December 1846 ...	20	44	75	14	1125	+ 4·68	W. 12·36
	Singi Meridional	Patángri H.S.	December 1861 ...	22	52	73	56	922	+ 0·70	W. 1·66
		Sáler „	March 1845 ...	20	43	73	59	5140	+ 2·26	„ 5·98
		Párnera „	February 1843 ...	20	33	72	59	614	+ 10·82	„ 28·86
		Kalsubai „	December 1842 ...	19	36	73	45	5400	+ 0·02	„ 0·06
	Káthiáwár Meridional	Dúngarpur H.S.	December 1852 ...	22	48	71	2	404	+ 2·29	W. 5·45
		Konkáwáo T.S.	October 1853 ...	21	39	70	59	622	+ 0·76	„ 1·91
	Gujarát Longitudinal	Sanoda T.S.	December 1851 ...	23	7	72	48	250	+ 3·75	W. 8·78
		Ingrori „	April 1852 ...	22	57	71	51	152	+ 1·18	„ 2·79
	Cutch Coast ...	Háthria H.S.	October 1856 ...	23	27	69	5	696	− 5·12	E. 11·80

Section	Series	Station		Date of Observation	Latitude North = φ		Longitude East of Greenwich		Height above Sea Level	Observed *minus* Computed Azimuth = (A − G)	(A − G) × cot φ
					°	′	°	′	*feet*	″	″
BURMA	Burma Coast*	Semu Tán	H.S.	January 1865 ...	22	49	91	50	226	− 7·36	E. 17·49
		Fi Tán	„	December 1865 ...	21	49	92	11	563	− 10·90	„ 27·23
		Dattaung	„	Nov. and Dec. 1866 ...	20	13	93	4	455	− 3·41	„ 9·26
		Taungzun	„	March 1884 ...	16	26	97	43	854	− 12·73	„ 43·16
		Southern Moscos	„	December 1877 ...	13	50	97	58	1186	− 10·14	„ 41·18
		Mergui Base-line, East End	T.S.	January 1882 ...	12	22	98	49	20	− 10·59	„ 48·30
		Mergui Base-line, W. End	T.S.	January 1882 ...	12	22	98	46	18	− 10·90	„ 49·71
		Natkalintaung	H.S.	December 1881 ...	12	26	98	46	888	− 10·79	„ 48·94
		Minthantaung	„	December 1881 ...	12	20	98	50	1054	− 11·43	„ 52·28
	Mandalay Meridional*	Myayabepgkyo	H.S.	November 1889 ...	18	22	96	25	1411	− 12·03	E. 36·23
		Toungoo	S.	February 1890 ...	18	56	96	28	186	− 15·79	„ 46·03
		Letpataung	H.S.	February 1891 ...	19	34	96	31	3975	− 16·99	„ 47·80
		Taungpila	„	March 1891 ...	20	42	95	56	1012	− 11·00	„ 29·11
		Mingun	„	February 1892 ...	22	3	96	2	1343	− 14·98	„ 36·98
		Sheinmaga	„	February 1892 ...	22	17	96	1	456	− 16·58	„ 40·46
		Malè	„	March 1892 ...	23	3	96	0	848	− 14·43	„ 33·91
		Ubyètaung	„	April 1894 ...	23	41	96	0	2766	− 11·58	„ 26·40
		Thonbinzin	„	February 1894 ...	24	14	96	1	1932	− 15·21	„ 33·79
		Seikpa	„	January 1895 ...	24	36	95	48	3857	− 19·02	„ 41·54
	Manipur Longitudinal*	Tamunja	H.S.	March 1896 ...	24	39	94	39	3387	− 6·76	E. 14·73
		Thyoliching	„	December 1898 ...	25	0	94	46	6566	− 8·68	„ 18·61
		Loijing	„	February 1899 ...	24	44	93	46	6635	− 9·36	„ 20·32

* The quantities entered against this Series are preliminary values.

NOTE:—H.S. signifies Hill Station (Principal).
S. „ Station (Principal) in the plains.
T.S. „ Tower Station (Principal).
Trs. S. „ Trestle Station (Principal).
P.S. „ Platform Station (Principal).